ABDULRAHMAN AL-SIRHA
JENS ERIKSEN & RICHARD

BIRDS OF THE
MIDDLE EAST

A PHOTOGRAPHIC GUIDE

H E L M

LONDON · OXFORD · NEW YORK · NEW DELHI · SYDNEY

This book is dedicated to BirdLife International and OSME

HELM
Bloomsbury Publishing Plc
50 Bedford Square, London, WC1B 3DP, UK
29 Earlsfort Terrace, Dublin 2, Ireland

BLOOMSBURY, HELM and the Diana logo are trademarks
of Bloomsbury Publishing Plc

First published in the United Kingdom 2022

A catalogue record for this book is available from the British Library.
Library of Congress Cataloguing-in-Publication data has been applied for.

ISBN: PB: 978-1-4729-8675-7; ePub: 978-1-4729-8676-4;
ePDF: 978-1-4729-9307-6

2 4 6 8 10 9 7 5 3 1

Design by Susan McIntyre
Printed and bound in China by RR Donnelley Asia Printing Solutions Limited Company

CONTENTS

FOREWORD

Birds are all around us. They are present on every continent, utilise almost every habitat on the planet, and most people around the world will encounter them on a daily basis. Birds can survive the freezing temperatures of the Antarctic, cross the highest of mountain ranges and the widest of seas, and endure the oppressive heat of the great deserts. But, as with so much of the wildlife and natural resources of our planet, birds are facing an uncertain future. To ensure their continued survival we need more people to appreciate the wonders of the avian world.

Over 800 different bird species make the Middle East their home for at least some part of the year. There are species that both breed and winter in various parts of the region, and others are just visitors during the winter months. The region connects the landmasses of Europe and Asia in the north to Africa in the south, and each year tens of millions of birds, large and small, pass through the region on their incredible migratory journeys. A small number of species are found only in this region and nowhere else on the planet. The more that people, both those that live in the region and those that visit, know about the birds around them the more likely they are to want to protect them and safeguard their futures.

This wonderful photographic guide makes a significant contribution to inspiring current and future generations to admire, understand and appreciate the wonderful birds of the Middle East. The high-quality photographs are provided by AbdulRahman Al-Sirhan and Jens Eriksen, who are lifelong advocates for the birds of the region. The photos, along with the concise species accounts, will enable the species most likely to be encountered to be identified and named with a high degree of certainty. However, as well as being able to name the species that you may encounter, knowing where to look for them is key. The section 'Good Birdwatching Sites in the Region' is a valuable description of many of the Middle East's top locations. A visit to any one of these sites will be an experience not to be forgotten.

This book provides a fantastic introduction to the birds of the region that will hopefully stimulate a lifelong interest in the avifauna of the Middle East and beyond.

Dr Rob Sheldon,
Chairman, Ornithological Society of
the Middle East, the Caucasus and
Central Asia (OSME)

INTRODUCTION

This photographic guide is aimed mainly at those adventurers who on their travels in the Middle East would like to spend some time watching the exciting birds that the region has to offer. We also hope it will encourage those who live in the region to take an interest in its wonderful birds and their conservation.

It is not comprehensive. For such a large region, which is home to over 800 species of birds, we have had to be selective in those we have chosen to include. In doing so we have been guided by the thought of what the enquiring visitor is most likely to see, rather than the specialities – though many of these are included too. While we hope that it will also provide pleasure and helpful information for serious birdwatchers, for these there are other, more detailed and comprehensive field guides that go into the finer points of plumage, voice and distribution that this dedicated audience requires.

Bird names, both English and scientific, are frequently changing and there is no one recognised authority, so for those interested in the boring bits, we have followed the taxonomy and nomenclature used by the International Ornithological Congress (IOC) as do most other bird books that cover the region (see page 220).

The area we cover ranges from the Iraq mountains in the north through the whole of Arabia and west to the countries of the Levant. So we include Bahrain, Iraq, Israel, Jordan, Kuwait, Lebanon, Oman, Palestine, Qatar, Saudi Arabia, Syria, the United Arab Emirates and Yemen (but not Socotra) – wonderful countries with a wonderful array of exciting habitats for any visitor to enjoy. Who can fail to be enthralled by a palm-fringed oasis, an acacia woodland in bloom, the sandy beaches, mudflats and mangroves of the Gulf, Red Sea and Arabian Sea coasts, Iraq's Mesopotamian Marshes or, of course, the ever-changing moods of the desert?

We have included a section on good sites to watch birds (pages 9–20), selecting a few of the 'must visit' localities that should be on the itinerary of the nature-loving tourist. In some of these accounts birds are mentioned that are not illustrated in this guide, but nonetheless we thought it helpful to include them for those with a keener interest in birds. Furthermore, in the section on useful books and contacts (page 220) there is more information, set out country by country, that will help to plan a visit and make it more rewarding.

Any visitor to the Middle East must be aware of the culture and religious sensitivities. Whilst some countries are very liberal in their ways, others are extremely conservative and their customs must always be respected, especially in the way you dress. So do make sure you are aware of these before you visit. Middle East countries are famous for their hospitality so always accept any kindnesses shown to you with grace. Always be friendly and never hold back about sharing with people your pleasure at watching their birds.

Finally, remember the birds you are enjoying. Their welfare must always come first. Don't go too close to nesting birds, particularly when parents are trying to feed their young. Remember that roosting and feeding birds, especially waders on shores and mudflats, are easily disturbed so take care not to approach too closely. There is always a temptation to get nearer and nearer to birds, especially when taking photographs, but try to stay at a sensible distance and give them the space they deserve.

Acknowledgements
For help with preparing 'Good Birdwatching Sites in the Region' we would like to thank Dan Alon (Israel), Nabegh Ghazal Asswad (Syria), Imad Atrash (Palestine), Oscar Campbell (UAE), Soumar Dakdouk (Lebanon), Laith El-Moghrabi (Jordan), Gavin Farnell (Qatar), Fouad Itani (Lebanon), Jonathan Meyrav (Israel), Hana Ahmad Raza (Iraq) and Assad Serhal (Lebanon).

Photographs used in this section were kindly provided by Mansur Al Fahad (Al Ha'er Wetland, Saudi Arabia), Issam Al-Hajjar (Syria), Abdulla Al Kaabi (Bahrain), Imad Atrash (Palestine), Samer Azar (Lebanon), Jem Babbington (Al Mahvar Tourist Park, Saudi Arabia), Oscar Campbell (UAE), Gavin Farnell (Qatar), Nashat Hamidan (Jordan), Jonathan Meyrav (Israel) and Hana Ahmad Raza (Qara Dagh, Iraq).

BIRD CONSERVATION

In a region where over 800 species of birds have been recorded, determining those for conservation intervention is challenging. Fortunately, the International Union for the Conservation of Nature (IUCN) and BirdLife International have developed strict scientific criteria placing the world's birds in the following categories: Extinct, Extinct in the Wild, Critically Endangered, Endangered, Vulnerable, Near Threatened, Least Concern, Data Deficient and Not Evaluated. These categories guide conservation priorities and actions. Top of the list of conservation priorities in the Middle East is the Critically Endangered Sociable Lapwing, which stops over in the region on its annual migrations but is, unfortunately, a target for hunters. Such illegal hunting is a serious conservation concern in the Middle East and tackling it is now part of a major BirdLife International programme, supported by OSME. The other Critically Endangered species in the region is Slender-billed Curlew, not illustrated and not recorded for nearly 40 years; indeed, it may be globally extinct.

Nine globally Endangered species occur, several of them very rare and most unlikely to be seen. One encountered in all countries in the Middle East is Egyptian

Vulture, the population of which has plummeted globally in recent years. In the Middle East, however, numbers are healthy. They are rarely subjected to deliberate poisoning at garbage dumps, nor have they succumbed to the veterinary drug, diclofenac, as has been the case elsewhere in its range. Fortunately, diclofenac, which is used for the treatment of cattle, is now being banned in Middle Eastern countries and replaced with safer alternatives not toxic to vultures when they feed on livestock carcasses. Another Endangered raptor is Steppe Eagle, which also congregates at garbage dumps in winter. Large soaring birds, such as the vulture and eagle, are especially vulnerable on migration, not only from shooting but also from electrocution by power lines and killing by wind turbines if these are inappropriately sited on their migration routes. Tackling these threats is now part of a multi-organisational effort – the Migratory Soaring Birds Programme – which is supported by the Global Environment Facility, the United Nations Development Program and BirdLife International.

One of the most Endangered songbirds in the region is Basra Reed Warbler, famous in the marshes of Iraq, virtually the only place it can be found nesting. A further 15 species occurring in the region covered by this book are categorised as globally Vulnerable and 31 as Near Threatened. Full details of all threatened species can be found on BirdLife's Data Zone (see page 220).

Bird conservation is not just about action for individual species. At its heart is the safeguarding of the environment and especially the places and habitats in which birds, indeed all wildlife and humans, live. In the Middle East an ambitious BirdLife project has identified all the Important Areas for Birds and Biodiversity – the IBAs. These 'jewels in the crown' are top priorities for protecting the region's wildlife. They are often a country's first step in developing a protected area network of nature reserves and national parks. Many are regularly monitored to assess threats and to ensure they still qualify under the strict criteria that IBA designation requires. As well as IBAs, the region's important wetlands have been identified and most countries have signed the Convention on Wetlands. This intergovernmental treaty, also known as the Ramsar Convention, provides the framework for national action and international cooperation for the conservation and wise use of wetlands and their resources. Two of the most important in the Middle East, the Mesopotamian Marshes of Iraq and Barr Al Hikman in Oman, are Ramsar wetlands and the Iraq Marshes are now a World Heritage Site.

Habitats, the places birds live, face many threats, and conservation bodies are active in tackling them. High on the list of concerns is the drainage of wetlands, overgrazing of natural grasslands by sheep and goats, the destruction of native woodlands and the destruction of the intertidal mudflats so important for wading birds to stop and feed on migration. Much of the conservation action involves political 'lobbying' – intervention at government and decision-making level to highlight the importance of sustainable land management and the damage that can be caused to these fragile habitats. Such damage not only impacts on wildlife, but also on the lives of people.

An overview of conservation is not complete without including the damage that pollution and the unregulated use of pesticides can cause to birds and our environment. Although the discharging of oil at sea is now banned, accidents happen, and the resulting pollution seriously impacts our marine wildlife. Fortunately, such incidents are rare in the Middle East. Less obvious is the insidious effects of pesticide poisoning, which occur when chemicals intended to control agricultural pests affect non-target organisms such as birds, other wildlife and humans. Accumulating pesticide residues in the air, water, soil, foods, plants and animals is a wildlife and human conservation issue yet to be seriously addressed in the Middle East.

A note on the photos
All images show the adult form, unless stated otherwise. Please refer to the key below for further information:

Adult – ad.	Breeding – br.	Spring – spr.
Juvenile – juv.	Non-breeding – non-br.	Summer – sum.
Immature – imm.	Winter – win.	

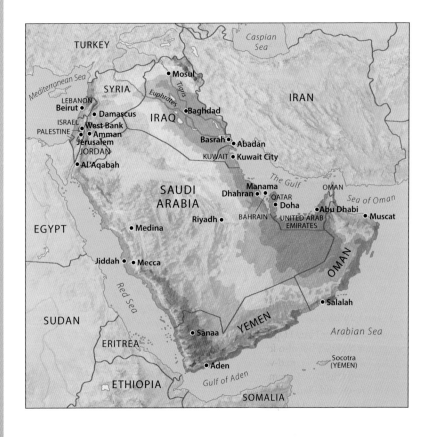

GOOD BIRDWATCHING SITES IN THE REGION

In this section we have chosen a small selection of sites that we feel would be of interest to a visitor who wants to spend some time birdwatching. Space does not allow it to be comprehensive but see page 220 for details of field guides and useful information to help plan your visit.

Note that in some of the accounts, notably those for Saudi Arabia and Yemen, a few of the very local species listed are not illustrated in this book but can be found in *Birds of the Middle East* (Porter & Aspinall, Helm, see page 220).

BAHRAIN

Dohat Arad Lagoon 26°15'N 50°37'E.
Shallow waters and scattered mangroves make this a good site for waders. Common birds include Terek, Marsh and Broad-billed Sandpipers, Lesser and Greater Sand Plovers, Pacific Golden Plover, Temminck's Stint, Pallas's, Lesser Black-backed and Slender-billed Gulls, Striated and Western Reef Herons, Crab-plover and Greater Flamingo.

Dohat Arad Lagoon.

Hamala Farm 26°08'59.4'N 50°27'57.9'E.
A site well known for Grey Hypocolius and Egyptian Nightjar, but also for Grey Francolin, Graceful Prinia, Red-vented and White-eared Bulbuls, Common Myna, Isabelline, Desert and Pied Wheatears, Namaqua Dove, Eurasian Hoopoe, Indian

Hamala Farm.

Silverbill, Siberian Stonechat, Pallid Swift, Blue-cheeked and European Bee-eaters.

Al Areen Wildlife Park and Reserve
26°00'N 50°29'E.

This site is a nature reserve and zoo. Namaqua and Laughing Doves, Grey Francolin and African Sacred Ibis (introduced), White-tailed and Northern Lapwings, Little Ringed Plover, Asian Desert Warbler and Grey Hypocolius can all be seen there.

IRAQ

Tourism is gradually developing in Iraq and the country holds many wildlife treasures for the enquiring ecotourist and birdwatcher. The following three sites are top of the list for any visit.

Mesopotamian Marshes

These vast, internationally famous wetlands in the south of the country are best approached from the town of Basra and comprise the Central Marshes, East Hammar, West Hammar and Hawizeh. Numerous wetland birds breed in the marshes, especially herons, waders and kingfishers (Pied Kingfisher is common). Specialities include Sacred Ibis, African Darter, Marbled Duck, White-tailed Plover, Iraq Babbler, Grey Hypocolius, Dead Sea Sparrow and Basra Reed Warbler. The marshes are virtually the only breeding area in the world for the last species. In winter they host tens of thousands of wildfowl, notably Marbled and Ferruginous Ducks, and many birds of prey. The marshes are best enjoyed by boat, an option that the local, friendly Marsh Arabs can help with.

The Mesopotamian Marshes.

Peramagroon Mountain 35°50′N 45°24′E.

The mountains of Peramagroon and Qara Dagh are relatively close to Sulaimani in Iraqi Kurdistan. Both offer a good selection of the region's special birds. Peramagroon is the highest mountain in the area and its rocky and oak-covered slopes have an exciting mix of breeding birds. Specialities include Egyptian Vulture, See-see Partridge, Upcher's Warbler, Eastern and Western Rock Nuthatches, White-throated Robin,

Qara Dagh.

Finsch's and Black-eared Wheatears, Cinereous Bunting and Masked Shrike.

Qara Dagh 35°18′N 45°23′E.

This is one of the most important sites in Iraqi Kurdistan for birds, mammals and plants. Special birds include Griffon and Egyptian Vultures, Golden Eagle, See-see Partridge, Eastern and Western Rock Nuthatches, White-throated Robin, Cinereous Bunting, Masked Shrike and Black-eared Wheatear. The Persian Leopard is still found here and there are plans to create a national park. Nature Iraq will soon have an ecolodge where visitors can stay (see page 220).

ISRAEL

Eilat 29°34′22.8′N 34°58′15.9′E.

Lying on the Red Sea coast and with impressive mountains to the west, Eilat is one of the best migration spots in the world. It is good all year, but especially from March to May. The Eilat Bird Watching Park (north of the city and near the Jordan border crossing), with its salt and freshwater habitats, is good for resident and migrant waders and passerines. It has a bird ringing scheme and is open for visitors. The Eilat Mountains are famous for watching migrating soaring birds, and in spring thousands of buzzards, eagles and storks pass over daily. Resident desert species include Hooded and

Eilat.

Hula Valley.

White-crowned Wheatears, Sand Partridge and Sinai Rosefinch. North of the city on route 90 are the vast K20 salt pans, which host thousands of waders, ducks, herons and gulls and a regular population of Greater Flamingos.

Hula Valley 33°06′51.1′N 35°35′40.2′E.

Close to Israel's northern border, this magnificent re-flooded wetland holds huge numbers of Common Crane, waterfowl, birds of prey and more. It is good all year, but best in winter. Up to 40,000 Common Cranes, along with Eastern Imperial and Greater Spotted Eagles and Pallid Harriers, winter in Agamon Hula Park. In the Hula Nature Reserve (Israel's first), famous for its historic papyrus reedbeds, there are Marbled Duck and a large harrier roost.

Jerusalem Bird Observatory
31°46′45.0′N 35°12′21.4′E.

This 'oasis' of urban nature in the middle of the city of Jerusalem is the only bird observatory in the world located on parliament grounds. It includes a small patch of natural habitat, a pool, bird hide and visitor centre. Open to the public all year, it offers a safe haven and is great for both migrants and resident birds.

Nizzana 30°50′59.3′N 34°27′08.1′E.

In the western Negev, at the end of road 211 near the Egyptian border, this semi-desert most notably is home to about 50 pairs of Macqueen's Bustards, which display from January to April, the best time to visit. Other species include Cream-coloured Courser, Spotted, Crowned and Pin-tailed Sandgrouse, Little Owl, and several species of larks and wheatears. Nizzana is also good for soaring birds and other migrants.

Ma'agan Michael
32°32′56.8′N 34°54′24.5′E.

On the Mediterranean coastal highway (Road 2) between Tel Aviv and Haifa, this area of fish ponds, reservoirs and beaches is good for birds all year. The fish ponds hold large numbers of herons, gulls, storks, ducks, waders and three species of kingfishers. Large numbers of White Pelicans occur on migration and rare birds include Pallas's Gull and Citrine Wagtail. A birdwatching park with trails is due to open soon.

Kfar Ruppin 32°27′27.2′N 35°33′46.9′E.

Near Beit Shean in the north Jordan Valley and located off Road 90 close to the King Hussein border crossing, this is an area of fish ponds, fields and natural habitats. Besides large numbers of herons, gulls, pelicans, waders, storks and birds of prey, the natural areas are good for Black Francolin, Desert Finch, Namaqua Dove and Blue-cheeked Bee-eater. In autumn it is one of the best places to see flocks of migrating White Storks and European Honey Buzzards. A birdwatching park with hides is due to open soon.

JORDAN

Azraq Oasis 31°50′N 36°49′E.

Once considered the most important stopover for waterbirds in the Levant, it currently hosts a sliver of the numbers it used to have 60 years ago. The remaining areas of permanent ponds and reedbeds are worth visiting during migration seasons, especially in spring. In good winter seasons, its mudflat (Qa') becomes flooded and it once again becomes the waterbird refuge that it once was, with a wide variety of ducks and waders. The main oasis has boardwalks and a bird hide.

Azraq Oasis.

Wadi Dana 30°40′N 35°35′E.

Located in the heart of Dana Biosphere Reserve, the wadi offers a wide variety of special birdwatching experiences. From the main viewpoint, several Middle Eastern specialities can easily be spotted, including Tristram's Starling and White-spectacled Bulbul. The same spot provides a raptor migration spectacle during the spring, with tens of thousands of Steppe Buzzards, Black Kites and several species of eagles. It is possible to hike down the wadi from Dana village all the way to Feynan in Wadi Araba. This walk passes through a variety of different

Wadi Dana.

habitats, from Mediterranean woodland to Saharo-Arabian gravel desert.

Aqaba 29°32′N 35°00′E.

Located at the northern tip of the Gulf of Aqaba, Aqaba is the main migration bottleneck for raptors in Jordan, especially during spring migration. Hundreds of thousands of more than 30 species of soaring birds pass over during spring and autumn migration. Aqaba Bird Observatory to the west of the town has a series of wastewater lagoons surrounded by strips of trees that attract a wide variety of waterbirds and passerines breaking their migration to rest and feed.

El-Harrah Desert 31°58′N 37°29′E.

This huge expanse of roughly 300km² of basalt rock desert lies between Azraq Oasis and the Iraqi border. It has two of Jordan's unique specialities: Basalt Wheatear and the dark morph Desert Lark known only from this desert. Several recently dug ponds (to provide water for the Bedouins and their livestock) have become resting places and drinking sources for resident desert species such as Thick-billed Lark and Black-bellied Sandgrouse, and also for wintering Steppe and Eastern Imperial Eagles. A four-wheel drive vehicle is required off the main highway.

KUWAIT

Jahra Pools Reserve 29°21'N 47°41'E.

This coastal wetland is close to Jahra town on Road 80, to the west of Kuwait City. Greater Spotted Eagles spend six months of the year here. Mountain Chiffchaff, Moustached Warbler, shrikes, three species of lapwings and three species of kingfishers can also be seen. Waders include Terek and Broad-billed Sandpipers, and Lesser and Greater Sand Plovers.

Jahra Farms.

Jahra Pools Reserve.

Sulaibiya Pivot Fields 29°15'N 47°45'E.

A large private farm irrigated by centre-pivot irrigation, this site is a wintering ground for eagles, buzzards, four species of lapwings, larks, wagtails, shrikes and warblers.

Al-Shaheed Park 29°22'N 47°59'E.

A large park in Kuwait City with pools and lawns, attracting many wintering and migrating birds, including shrikes, buntings and warblers. The saline depression with a small reedbed, in the northern corner, is the best place to start. Fintas Park and Green Island are other good sites in Kuwait.

Jahra Farms 29°21'N 47°40'E.

These traditional farms with green fields in the centre of Jahara City are a good site for thrushes, pipits, shrikes, stonechats, flycatchers, mynas and White-eared Bulbul.

LEBANON

Aammiq Wetland Reserve 33°43'N 35°47'E.

The top birdwatching site and largest remaining freshwater wetland in Lebanon, this reserve is surrounded by areas of rough grazing, cultivation and trees, all of which add to its rich biodiversity. Breeding birds include Little and Great Crested Grebes, Little Bittern, Western Marsh Harrier, Water Rail, Calandra Lark, Penduline Tit and Black-headed Bunting. It is also very important for migrant and wintering birds. In addition to the thousands of raptors, herons, storks, pelicans and cranes, specialities include Marbled and Ferruginous Ducks, Greater Spotted and Eastern Imperial Eagles, Pallid Harrier and Syrian Serin.

Aammiq Wetland Reserve.

Ras Baalbek semi-desert 34°17'N 36°22'E.

In the northern Beqaa region this dry, barren area hosts many species rarely found elsewhere in Lebanon. Residents include Long-legged Buzzard, Little Owl, Temminck's Lark, Western Rock Nuthatch, Streaked Scrub and Spectacled Warblers, Mourning Wheatear and Trumpeter Finch. In summer you can find Eurasian Stone-curlew, Cream-coloured Courser, Bar-tailed, Lesser Short-toed and Desert Larks, and Pale Rock Finch. On migration Eurasian Dotterel, Northern Lapwing and Finsch's Wheatear also occur.

Anjar-K'far Zabad 33°44'N 35°56'E.

Located in the eastern Beqaa Valley, at the foot of the Anti-Lebanon mountain range, this site consists of farmland, freshwater wetland, mixed woodland and open scrubby hillsides. The historic ruins of Anjar are a breeding site for Syrian Serin. Other breeding birds include White Stork, Eurasian Hoopoe, Western Rock Nuthatch and Cretzschmar's Bunting as well as several species of shrikes, wheatears and warblers. The reedbeds hold good numbers of Black-crowned Night Herons, Little Bitterns and Penduline Tits.

Al Shouf Cedar Nature Reserve 33°45'N 35°43'E.

Located on the slopes of Barouk Mountain, this is Lebanon's largest nature reserve, stretching from Dahr Al-Baidar in the north to Mount Niha in the south. The oak, juniper and magnificent cedar forests are a home and stop-over site for over 200 species, including Chukar, Golden Eagle, Griffon Vulture, Turtle Dove, Eurasian Jay, Common Raven, Horned Lark, Woodlark, Common Redstart, Rock Sparrow, Rock Bunting and Syrian Serin.

Horsh Ehden Nature Reserve
34°17'46.6'N 35°58'18.2'E.

On the north-western slopes of Mount Lebanon, with its beautiful cedar, juniper and fir forest, this reserve offers a great opportunity to watch the impressive soaring-bird migration of storks, pelicans, eagles and buzzards. During autumn, visit the ABCL observatory (34°17'58.6"N 35°58'36.3"E) for a chance to participate in counts of soaring birds. Other species here include Bonelli's Eagle, Western Rock Nuthatch and Syrian Woodpecker.

OMAN

As Sayh, Musandam Peninsula
25°59'N 56°13'E.

This site, at an altitude of 1,200m, lies along the graded road from Khasab south towards Jabal Harim and Wadi Bih. It is an excellent spring migration site, especially for shrikes (nine species have been recorded), wheatears (10 species), buntings and vagrants. April is the best month to visit.

As Sayh.

Al Ansab Wetland 23°34'N 58°20'E.

A series of wastewater treatment lagoons in Muscat. From the main highway from the airport take the exit for the Oman Exhibition and Convention Centre and continue to the treatment plant. This location provides an excellent introduction to birdwatching in Oman with a good variety of ducks (including Ferruginous Duck), herons, waders and rails. It is best visited during the cooler months. There are trails and hides. Reservation is free but essential (see page 220).

Barr Al Hikman 20°42'N 58°42'E.

These extensive tidal mudflats on the east coast of mainland Oman, opposite Masirah Island, are probably the most important site for migratory and wintering shorebirds in the Middle East. Hundreds of thousands of waders can be seen from September to May. Highlights are large numbers of Greater Flamingo (10,000), Great Knot (up to 1,000), Broad-billed Sandpiper (hundreds), Bar-tailed Godwit (80,000) and Crab-plover (8,000).

Ayn Hamran 17°06'N 54°17'E.

This excellent site comprises a park with lush vegetation around a permanent spring in the hills 13km east of Salalah. It is well signposted. Ayn Hamran has all the special passerines, owls and eagles of southern Oman. Highlights are Arabian Scops Owl, Arabian Warbler, Palestine and Shining Sunbirds, Black-crowned Tchagra and Arabian Golden-winged Grosbeak.

Khawr Rawri 17°02'N 54°26'E.

A large coastal lagoon about 35km east of Salalah, this location is well signposted as it is also an archaeological site (Sumhumran). One of Oman's best birdwatching locations, it has an excellent variety of waterbirds and landbirds. Highlights include Arabian Partridge, Cotton Pygmy Goose, Pheasant-tailed Jacana, Black Stork, and Steppe, Eastern Imperial and Greater Spotted Eagles.

Wadi Darbat.

Wadi Darbat 17°06'N 54°27'E.

A valley (wadi) with permanent water in the Dhofar Mountains just north of Khawr Rawri, Wadi Darbat has an interesting variety of migratory and resident birds. Common species include Arabian Partridge, Eastern Imperial Eagle, Bruce's Green Pigeon, Grey-headed Kingfisher, African Paradise Flycatcher, Blackstart, Abyssinian White-eye, Rüppell's Weaver, Cinnamon-breasted Bunting, Tristram's Starling and Fan-tailed Raven.

PALESTINE

Jenin 31°42'N 35°19'E.

Situated at the foot of the rugged northern-most hills (Jabal Nablus) of the West Bank, Jenin is one of the most important sites in Palestine for biodiversity. The national flower of Palestine, *Iris haynei*, grows here. In spring and autumn, the

Jenin.

Judean Desert.

Shamal Coast.

Jenin district is one of the most important localities for observing the migration of soaring birds, notably White Stork, Common Crane, White Pelican and Lesser Spotted Eagle. The wet plains also attract many migrant birds in winter, including birds of prey. The waste dump in Zahrat Al-Finjan attracts Black Kites, vultures, storks and gulls.

Jerusalem Wilderness Site (Judean Desert)
32°21'N 35°16'E.

With breathtaking scenery and unforgettable views over the Dead Sea, this is an excellent location for watching many species of migrating birds of prey, notably Egyptian Vulture, Levant Sparrowhawk, Lesser Spotted Eagle and European Honey Buzzard. The biodiversity doesn't end with birds; interesting mammals include Rock Hyrax, Syrian Jackal, Red Fox and Striped Hyena, as well as many lizards and beetles.

QATAR

Shamal Coast 26°08'39.6'N 51°16'10.5'E.
A bottleneck in spring for migrant birds funnelling up the Qatar peninsula and a resting place for tired autumn birds that have crossed the Arabian Gulf. This is probably the best place in Qatar to see Crab-plover and Cream-coloured Courser (August to September are the best months) one can expect all manner of waders, gulls and terns throughout the year. The coastal path alongside the mangroves to the east of the town often turns up good birds, including many passerines. Nearby Shamal Park (26°07'28.8'N 51°12'38.0'E) is always worth a visit. Its lawns, gardens and maintained trees and bushes cover 2.5ha.

Qatar 02 MIA Park and Al Bidda Park
25°17'39.1'N 51°32'36.7'E and 25°17'58.4'N 51°31'03.7'E.

These two coastal parks in the heart of Doha are often overlooked by local birders, but regularly turn up good birds and are a great place for the short-stay visitor to watch birds. Expect wagtails, pipits, shrikes and buntings during migration. The parks consist of many hectares of lawns, with pockets of shrubs, and pathways. Long lenses (500mm+) can draw unnecessary attention and require special permission.

SAUDI ARABIA

Al Ha'er Wetland 24°22'N 46°54'E.
This is also called the Riyadh River, which runs from Riyadh Sewage Treatment Plant for about 50km. A good starting point is at the coordinates above. It is a good a site for wetland birds, including Purple Heron, Little Bittern,

Al Ha'er Wetland.

Al Mahvar Tourist Park.

Black-crowned Night Heron, Great Reed, European Reed and Moustached Warblers. Spur-winged and White-tailed Lapwings are also common. Raptors are represented by Greater Spotted, Eastern Imperial and Steppe Eagles and Pallid Harrier in winter and on migration. Also, Arabian Green and Blue-cheeked Bee-eaters are common.

Haradh 24°9'N 48°54'E.

Haradh is an industrial city in Ahsa Governorate with a large number of pivot irrigation fields. Sociable Lapwing is now seen every winter together with Spur-winged and Northern Lapwings. Haradh is also excellent for raptors such as Pallid Harrier and Greater Spotted, Steppe and Eastern Imperial Eagles. Larks may occur in their thousands, with Greater and Lesser Short-toed Larks the commonest and Greater Hoopoe-Lark regular. The nearby large escarpment is good for Desert and Bar-tailed Desert Larks and Black-crowned Sparrow-Lark, as well as Cream-coloured Courser.

Al Mahvar Tourist Park 18°57'N 42°07'E.

The best site to see Arabian endemics. These include the Arabian Woodpecker, Arabian and Philby's Partridges, Arabian Scops Owl, Arabian Wheatear, Arabian Serin, Yemen Serin, Yemen Thrush, Red-breasted Wheatear, Yemen Linnet

and Yemen Warbler. Some otherwise mainly African species such as Arabian Eagle-Owl, Little Rock Thrush, Brown Woodland-Warbler, Abyssinian Nightjar, Abyssinian White-eye and Diedrik Cuckoo are also found.

Al Sadd Lake 18°12'N 42°29'E.

This location features a lake surrounded by trees and bushes. The endemic Arabian Waxbill can be seen, together with Helmeted Guineafowl, White-browed Coucal, Abdim's Stork, Pink-backed Pelican, Hamerkop, Gabar Goshawk, Dark Chanting-Goshawk, Grey-headed Kingfisher, White-throated and Blue-cheeked Bee-eaters, Abyssinian Roller, Black-crowned Sparrow-Lark, Singing Bushlark, Zitting Cisticola, Eastern Olivaceous and Upcher's Warblers, Arabian Babbler, Black Scrub Robin, Nile Valley Sunbird, Rüppell's Weaver, African Silverbill, Goliath and Striated Herons and African Grey Hornbill.

Tabuk Northern Farms 28°36'N 36°24'E.

These centre-pivot irrigation farms are mostly fenced but access can be granted from the owners or they can be birded from outside. Sociable Lapwing winters annually now, and resident birds include Spur-winged Lapwing, Black-winged Stilt, Desert Finch, Desert Lark, Sinai Rosefinch and Brown-necked Raven.

SYRIA

Jabboul 36°05′N 37°48′E.

A 40km drive south-east of Aleppo, the Jabboul shallow brackish wetland is the most important wetland in Syria. In winter it hosts more than 120,000 birds of 70 species including over 22,000 Greater Flamingos (there are about 4,000 breeding pairs) and 20,000 Eurasian Coots as well as many species of geese, ducks, waders, gulls and terns. Dead Sea Sparrow also occurs. Among the threatened species are White-headed and Marbled Ducks, also Sociable Lapwings during the spring on the adjacent steppe.

Coastal mountains 35°62′N 36°10′E.

This bottleneck mountain range, which concentrates soaring birds on migration, rises to 1,000–1,500m above sea level. There are several types of humid and sub-humid woodlands and maquis overlooking the steep slopes of the Al-Ghab Valley, which is considered the

Jabboul.

Coastal mountains.

Abu Zad and west of Damascus mountains.

northern end of the Rift Valley. Sites along the ridge are very good for watching thousands of birds of prey, storks and pelicans on spring and autumn migration.

Abu Zad and west of Damascus mountains 33°72′N 36°12′E.

This elevated area at the southern end of the Anti-Lebanon mountain range is a 50km drive from Damascus. Snow-covered in winter it comprises open semi-arid highland steppe with sparse grassland, small orchards, vertical cliffs and steep rocky slopes. It is located on the flyway for migrating soaring birds and is also home for several passerines, including Syrian Serin, Western Rock Nuthatch and Alpine Accentor, as well as rock thrushes, redstarts, larks and warblers.

UNITED ARAB EMIRATES

Abu Dhabi Island 24°28′N 54°22′E.

The island has urban parks, golf courses, fringing coastline, intertidal mudflats and mangroves. Migrant species such as shrikes and wheatears occur in spring and autumn. Western Marsh Harrier and Crested Honey Buzzard winter annually in small numbers. Greater Flamingos and Western Reef Egrets may be seen along shorelines. White-cheeked, Saunders's

Abu Dhabi Island.

Jebel Hafit, Al Ain.

and Bridled Terns, along with Osprey, breed on offshore islets. See page 220 for precise site details.

Al Wathba Wetland Reserve
24°15'N 54°35'E.

A Ramsar site 20km south-east of Abu Dhabi Island. Greater Flamingos have bred since 2011 and are easily observed at close range. Up to 5,000 shorebirds and waterfowl winter annually; more than 50 Western Marsh Harriers roost each evening in winter with Great Spotted Eagle present most winters. The reserve is open from 08:00, Thursdays and Saturdays, between November and April. There are trails, hides and visitor facilities.

Jebel Hafit, Al Ain 24°03'N 55°46'E.

This spectacular limestone mountain reaches an altitude of 1,240m. Recently declared a national park, it is immediately south-east of Al Ain, an oasis city with many tourist sites 130km east of Abu Dhabi Island. Jebel Hafit has breeding species typical of eastern Arabian montane habitats, including Hume's Wheatear and Desert Lark (both common), Sand Partridge and Lichtenstein's Sandgrouse. It is the only UAE site for Egyptian Vulture, which still occurs in small numbers. Persian

Wheatear and Plain Leaf Warbler winter from October to March. Green Mubazzarah is a parkland area with springs at the foot of the mountain; a spectacular tarred road ascends Jebel Hafit from nearby.

Ra's al-Khor, Dubai 25°11'N 55°19'E.

Intertidal mudflats and mangrove plantations on the edge of Dubai Creek. There are large overwintering populations of shorebirds, herons and Greater Flamingo. More than 20 Greater Spotted Eagles, along with Ospreys and Western Marsh Harriers, occur in winter; they are best observed as they start to thermal from 09:00–11:00. There are several viewpoints with hides, although navigating busy highways is necessary to get from one to another.

Al Mamzar Park, Sharjah
25°19'N 55°20'E.

A 'green lung' along the coastline of urban Sharjah, 10km from central Dubai. It is famous for rarities and migrants, including many warblers, shrikes, redstarts and flycatchers in spring and autumn. Purple Sunbird, Eurasian Hoopoe and Indian Roller all breed commonly and are easy to see. The park is open from 08:00 daily.

Khor al-Beida, Umma al Qwain
25°31'N 55°34'E.

Mainland UAE's most bird-rich intertidal wetland, lying north of the town of Umm al Qwainn (40km north of Dubai). Very large numbers of migrant and wintering shorebirds, with Crab-plover a local speciality; small numbers of Great Knot also occur. A large colony of Socotra Cormorant breeds in winter on Siniyia Island, which shelters the site from the Arabian Gulf.

Khor Kalba, Sharjah 25°00'N 56°21'E.

A beautiful narrow creek on the Gulf of Oman, lined extensively by old, gnarled mangroves, in the far south-east of the UAE. It is the only UAE site for Collared Kingfisher of the endemic subspecies *kalbaensis*. These are best seen at low tide, as are small numbers of shorebirds and terns (the latter mainly in the nearby harbour). Small numbers of breeding Sykes's Warbler, the only UAE population, breed in the mangroves and are best located when singing, from March to May. Indian Pond Heron winters regularly. Many interesting seabirds have been recorded from boat trips into adjacent waters.

YEMEN

Because of the tragic conflict in Yemen, this account is limited since visiting this wonderful country to watch birds is either not possible or unsafe. We hope that this will soon change as the mountains of Yemen are very special, holding many species that are endemic to south-west Arabia. Visiting any highland area, such as the Ibb Mountains (35°33'N 45°29'E), between Sana'a and Taizz, or Kawkaban (35°33'N 45°29'E), just an hour from Sana'a, will be rewarding as the wooded, agricultural terraces hold many of the south-west Arabian endemic species, including Philby's and Arabian Partridges, Arabian Woodpecker, Yemen Warbler, Yemen Thrush, Arabian Wheatear, Arabian Accentor, Arabian and Yemen Serins and Yemen Linnet.

Terraces near Ibb

The Red Sea and Arabian Sea coasts provide good opportunities to watch seabirds, especially gulls and terns, and many waders. Amongst the most notable species are Pallas's and Slender-billed Gulls, Lesser Crested and Greater Crested Terns, Greater and Lesser Flamingos and Crab-plover. Chestnut-bellied Sandgrouse and Black-crowned Sparrow-Lark are common on the neighbouring Tihama coastal plains.

Khor Kalba, Sharjah.

Yemen Terraces.

Chukar Partridge *Alectoris chukar* 33cm

A chicken-sized bird, very like an Arabian Partridge in plumage, but with a different face pattern and bolder stripes on the sides of the body. The male and female are similar in appearance. This partridge is more often heard than seen – a characteristic, chuckling *chukara-chukar, chukara-chukar*, which can carry a long distance.

Where to see Rocky and stony hillsides in those countries bordering the Mediterranean coast, as well as in Musandam, Oman and neighbouring mountains of UAE.

Arabian Partridge

Alectoris melanocephala 36cm

A chicken-sized bird that is similar in pattern to Chukar Partridge but always told by the black crown (grey in Chukar) and the fine barring on the sides of the body. Often heard calling, a far-carrying *kok, kok, kok, kok, kok, chok-chok-chok chook*, which accelerates and descends.

Where to see Dry, rocky and stony hillsides with bushes in Oman and parts of southern Arabia.

See-see Partridge

Ammoperdix griseogularis 24cm

Much smaller than the Chukar or Arabian Partridge and similar to Sand Partridge, but with a different distribution. Note the head pattern of the male and the rusty stripes that sweep down the sides of the body. The female is a rather featureless sandy grey. Runs fast when disturbed or flies low with whirring wingbeats. Has characteristic, far-carrying, repeated *who-wit* call.

Where to see Dry, stony hillsides and barren areas in Iraqi Kurdistan. Introduced on Sir Bani Yas island, UAE.

Sand Partridge

Ammoperdix heyi 24cm

Similar in size and plumage to See-see Partridge, especially the female, but sandier in coloration. The male has a white patch behind the eye, which is not bordered by black, and bolder stripes on the sides of the body. Calls a monotonous *qwei-qwei* … from a prominent look-out.

Where to see Rocky hillsides and sandy wadis in the UAE, Oman and southern and western Arabia – a completely different range to See-see Partridge.

Grey Francolin *Francolinus pondicerianus* 30cm

A greyish-brown, finely barred and vermiculated gamebird with no real distinguishing features apart from a black necklace on the male; slightly smaller than Chukar Partridge. Can often be seen in small parties in the open and is not secretive like other gamebirds. Has a loud, far-carrying call of up to 15 notes.

Where to see Scrubby areas and margins of fields close to the Gulf and in northern Oman.

Partridges

Common Quail *Coturnix coturnix* 17cm

The smallest gamebird in the region, just the size of a Common Starling. Rarely seen on the ground, where it creeps quietly through low vegetation. More often observed when flushed, and in flight note the plain, pointed wings, striped back and fast wingbeats.

Where to see In cereal fields, meadows and grassland almost anywhere in the region, mainly in spring and autumn when on migration.

Greylag Goose *Anser anser* 80cm

Probably the commonest goose to be seen in winter (but Greater White-fronted Goose *Anser albifrons*, not illustrated, identified by white surrounding the bill and black bars on the belly, can be fairly common in Oman). Note its large size, heavy pink bill and pink legs. Loud call, uttered in flight, is like that of domestic goose. Usually occurs in flocks.

Where to see Fields, marshes and estuaries throughout the region, except in much of Arabia.

Common Shelduck *Tadorna tadorna* 63cm

A large duck, the size of a small goose. The white body, green head, red bill and broad chestnut band around the breast and across the back make it unmistakable. Male and female are similar in plumage.

Where to see Coasts, marshes and wetlands in spring, autumn and winter throughout the region, though it is uncommon or rare in much of Arabia.

Ruddy Shelduck *Tadorna ferruginea* 64cm

Similar in size to Common Shelduck, but the orange-chestnut body and paler head make this duck unmistakable. Sexes similar, but in the breeding season the male has a narrow black neck-collar. In flight note the conspicuous large white wing patches.

Where to see Similar wetlands to Common Shelduck, but also in fields throughout the region, mainly in spring, autumn and winter, though rare in southern Arabia.

Cotton Pygmy Goose *Nettapus coromandelianus* 33cm

This rare visitor from Asia is the smallest duck to occur in the region. The male is distinct with a white head and neck, and a black cap, eyes and band across the breast. Female is less distinct, but both sexes have stubby goose-like bill. In flight shows conspicuous white trailing edge to the wing.

Where to see A bird of well-vegetated lakes, it is a rare winter visitor mostly to Oman, but vagrants have occurred throughout the region.

Garganey *Spatula querquedula* 39cm

One of the smallest ducks. The male is easily distinguished by the noticeable long white stripe on a chocolate head and blue forewing in flight. Female very similar to Eurasian Teal, but with longer bill and white and dark stripes on head. Like the other dabbling ducks, it doesn't dive. Often occurs in large flocks.

Where to see Can occur on almost any wetland throughout the region on spring and autumn migration.

♂

♂ ♀

Geese and ducks

Northern Shoveler

Spatula clypeata 51cm

The huge spatulate bill readily identifies both sexes of this dabbling duck. The male's plumage is distinctive with a green head, white breast and chestnut flanks and belly. The female is drab brown. In flight shows a blue forewing.

Where to see A migrant and winter visitor that can be seen on any wetlands throughout the region from autumn through to spring.

Gadwall

Mareca strepera 51cm

Very similar to the much commoner Mallard, only the male is easy to distinguish with its dark grey plumage, black surround to the tail and black bill. Both male and female can be identified in flight by a white patch on the hindwing, close to the body.

Where to see A migrant and winter visitor that can be seen on any wetlands throughout the region from autumn to spring.

Eurasian Wigeon
Mareca penelope 48cm

The rufous head with a yellow forehead and grey body with a black tail-end distinguishes the male from all other ducks. The female has a rather rufous-grey plumage, but both sexes show white belly in flight. Often seen grazing on short grasses at the edge of a wetland.

Where to see Migrant and winter visitor to coastal marshes and freshwater wetlands throughout the region from autumn to spring.

Mallard *Anas platyrhynchos* 56cm

The green head, brown breast, white neck-ring and yellow bill make the male of this large duck distinctive, but the brown female can appear very similar to the females of other 'dabbling' ducks – ducks that do not dive.

Where to see Although it breeds in a few places in the region, it occurs mostly as a migrant and winter visitor to coastal marshes and freshwater wetlands from autumn to spring.

Northern Pintail

Anas acuta 56cm (plus long tail of male)

Look for the long 'pin-tail' of the male, which is not always easy to see. The best feature to look for is the white stripe up the chocolate neck and head of the male. As with all dabbling ducks the female is brown with few notable features, but she will usually associate with a male.

Where to see Migrant and winter visitor to coastal marshes and wetlands throughout the region from autumn to spring.

Eurasian Teal

Anas crecca 36cm

The smallest of the common ducks seen in the region. The male is easily identified by its reddish-brown head, green mask through the eye, white line along the greyish body and yellow patches at the side of the tail. The brown female will often associate with her male partner, aiding identification.

Where to see Migrant and winter visitor to coastal marshes and wetlands throughout the region from autumn to spring.

Red-crested Pochard *Netta rufina* 56cm

The male of this colourful diving duck has a dark orange head and bright red bill, making it very easy to identify. The female is less distinct, but note the pale grey cheeks contrasting with a dark crown.

Where to see Mainly a winter visitor to lakes and wetlands in the north of the region, Mediterranean areas and the Gulf countries south to Oman.

Common Pochard *Aythya farina* 45cm

The world population of this diving duck is declining, and it is considered Vulnerable globally. The male is distinguished by its chestnut head and neck and grey body. The female is much duller, being a mix of rather drab browns and greys, but note the similar shape of the head and bill to those of the male.

Where to see Migrant and winter visitor to freshwater wetlands throughout the region from autumn to spring.

Ferruginous Duck *Aythya nyroca* 40cm

This diving duck is globally Near Threatened. Very uncommon in the region and only seen in small numbers. Slightly smaller than Tufted Duck with a high crown and sloping forehead. The male is a rich chestnut with a white eye and white undertail-coverts. The female is browner with a dark eye.

Where to see Although it breeds at a few wetlands in the region it is most likely to be seen in winter.

Tufted Duck

Aythya fuligula 42cm

The male of this diving duck can be instantly distinguished by its black-and-white plumage and, when close, a crest on the back of its head. The dark brown female and immature plumages lack white sides to the body but note the blue-grey bill with a dark tip.

Where to see Migrant and winter visitor to freshwater wetlands throughout the region from autumn to spring.

European Nightjar

Caprimulgus europaeus 26cm

A brown, grey and buff-white camouflage-patterned bird, rarely seen during the day. If it is seen, it will be because it has been flushed from its roost where it sits tightly on the ground or a low branch. In its wavering flight, note the long wings, with a white spot near tip in the male.

Where to see Any open areas with scattered vegetation on migration in spring and autumn.

Egyptian Nightjar *Caprimulgus aegyptius* 25cm

Similar to European Nightjar but much paler, being basically sandy-grey with feathers finely vermiculated with buff and black. In flight dark wings, which lack white spots, contrast with the rest of the pale plumage. Usually only seen when flushed.

Where to see A bird of semi-deserts, often with palms and scrub, both in its breeding areas in Iraq and the northern Gulf region, and on migration in spring and autumn throughout the region, though rare.

Alpine Swift

Tachymarptis melba
21cm; wingspan 57cm

The largest swift in the region and
quite distinct with its long, scythe-like
wings, brown upperparts and white
underparts with a dark band across
the breast. Flocks often heard calling,
a loud chattering trill, as they fly
high overhead.

Where to see Breeds in scattered
colonies in rocky mountains
in countries bordering the
Mediterranean, and in Iraq and
Arabia. Otherwise, can be seen
fairly widely on spring and
autumn migration.

Common Swift

Apus apus 16cm; wingspan 45cm

A dark brown swift with a whitish
throat, much smaller than Alpine
Swift but with similar long, scythe-
like wings and fast, sweeping flight,
often in groups high in the sky. Listen
for its high-pitched screeching call.

Where to see Breeds in countries
bordering the Mediterranean and in
parts of Iraq, usually nesting under
the eaves of buildings or on cliffs.
Otherwise widely seen during spring
and autumn migration.

Pallid Swift

Apus pallidus 16cm; wingspan 44cm

Very similar to Common Swift in size and shape, but rather paler and difficult to distinguish unless seen well. Look for the slightly broader head, larger whitish throat-patch, paler forehead and dark eye-patch. Note also pale scaling below and darker 'saddle' on back.

Where to see A summer visitor to scattered breeding colonies, mainly on cliffs, throughout the region. Also occurs widely on migration in spring and autumn.

Little Swift

Apus affinis 12cm; wingspan 34cm

The smallest swift in the region, rather stout with a short, square-ended tail. Note especially the conspicuous white rump. Somewhat similar to Common House Martin (see page 161) but is all-dark below (not white).

Where to see A summer visitor to isolated colonies mainly in southern Arabia, Iraq and countries bordering the Mediterranean. Nests colonially in caves and buildings. Migrating birds are mostly seen in Oman and eastern Arabia.

Macqueen's Bustard *Chlamydotis macqueenii* 60cm

Ground-dwelling bird, twice the size of a domestic chicken. Rather secretive, it holds a special place in Arab culture being a quarry for hunters with falcons. Quickly distinguished from other bustards in the region by the black frill down the side of the neck.

Where to see Breeds erratically in the region and recently introduced to several areas. Migrant and winter visitor to deserts and steppes in the region.

Common Cuckoo
Cuculus canorus 33cm

Can resemble a small falcon with its long wings, long tail and rather fast, shallow wingbeats. Grey or rufous-brown in colour, and always with finely barred underparts. Its characteristic *cuc-koo* song is normally heard only where it is breeding.

Where to see Breeds occasionally in areas on the Mediterranean coast; otherwise seen on spring and autumn migration throughout the region in areas with trees, especially those with caterpillars.

juv.

Chestnut-bellied Sandgrouse *Pterocles exustus* 32cm

One of six sandgrouse species in the region, all rather similar. Resembles a ground-dwelling pigeon in shape and size, but with a pointed tail. Like all sandgrouse is adapted to life in deserts, with absorbent breast feathers enabling water to be carried long distances to thirsty young. Often in flocks. In flight, note dark belly and dark underwings.

Where to see Resident in semi-deserts and deserts in UAE, Oman and parts of western Arabia.

Crowned Sandgrouse *Pterocles coronatus* 28cm

A rather pale, plain sandgrouse with a short tail distinguishing it from the rather similar Spotted Sandgrouse (*Pterocles senegallus,* not illustrated), which has a long, pointed tail. The female is sandy coloured but the male has an orange-and-grey head with a black face mask. Fast flying and usually occurs in flocks.

Where to see Resident of stony semi-deserts and deserts in Oman, UAE, Jordan and Israel, with isolated colonies in western Arabia.

Lichtenstein's Sandgrouse *Pterocles lichtensteinii* 25cm

Sandgrouse gather to drink at watering spots and for this species such gatherings occur just after dark, so it can be hard to observe. Small sandgrouse with square-ended tail and finely barred black and white on a yellowish-buff plumage. Note the breast-bars and head pattern of the male.

Where to see Rocky deserts and arid mountains with scrub in countries bordering the Red Sea and Arabian Sea, as well as Oman and eastern UAE.

Rock Dove *Columba livia* 33cm

A fairly familiar dove in the region, especially in its feral form when it can be found in small flocks in towns, in a variety of colour forms. The true Rock Dove is pale grey with two black bands on the wing and a white rump. Note its characteristic cooing song, *croo-oo-u*.

Where to see Rocky mountains, wadis and sea cliffs in much of the region, except where desert.

Common Wood Pigeon *Columba palumbus* 41cm

The largest pigeon to occur in the region and easily identified by the white patch on the sides of its neck and the white band across the wing when seen in flight.

Where to see Isolated breeding in parts of Iraq, Syria, Kuwait and Oman, otherwise an uncommon winter visitor mainly to the countries bordering the Mediterranean. A bird of woodland when breeding, but also fields in winter.

European Turtle Dove *Streptopelia turtur* 27cm

In the last 20 years this delightful dove has become a globally threatened species with numbers declining dramatically throughout its range. Note the rusty-edged feathers on dark-spotted upperparts and black and white streaked patch on the neck. Its song is a soft, deep purring.

Where to see A summer visitor that breeds rather patchily in lightly wooded areas throughout the region; widespread on spring and autumn migration.

Eurasian Collared Dove *Streptopelia decaocto* 33cm

Readily identified by its sandy-brown plumage with a black half-collar on the hind-neck, this is a fairly common dove in many towns and villages. Its three-note song, *co-coo-co*, is frequently heard.

Where to see Occurs in towns, villages and parks in much of the region, except for central and southern Arabia, where it is largely replaced by the very similar African Collared Dove (*Streptopelia roseogrisea,* not illustrated).

Laughing Dove

Spilopelia senegalensis 26cm

A common dove, which is
expanding its range throughout
the region. Often in pairs or small
groups, note its small size, dark
red-brown to sandy-brown plumage
with a black-spotted patch on the
neck-sides. Its song is a soft, five-
note cooing, from which it gets
its name.

Where to see Resident in towns,
villages, oases and fields in much of
the region, except for central Arabia.

Namaqua Dove *Oena capensis* 29cm (including 9cm tail)

The smallest dove in the region, recalling a large budgerigar in shape and size and often seen on the ground. Greyish plumage with a long, black tail and rufous wings when seen in flight. The male has a black face and upper breast.

Where to see A bird of semi-desert with scrub, patchily distributed in Arabia where commonest in the south and west. Range is expanding northwards.

Bruce's Green Pigeon

Treron waalia 31cm

The only green and yellow dove in the Middle East, and not easily seen when perched in the shade of a tree. Its song, a loud, querulous whistle, is not unlike the call of a Tawny Owl.

Where to see A resident of areas with tall fruit-bearing trees, especially figs, including palm groves on mountain slopes in south-west Arabia and southern Oman.

Water Rail *Rallus aquaticus* 26cm

Very secretive as it stalks through reeds and aquatic vegetation. Much smaller than Common Moorhen and its long red bill is distinctive. Note the adult's blue-grey underparts and the black-and-white bars on its sides. Young birds are duller.

Where to see Ponds and ditches with reeds and dense vegetation. Breeds patchily in a few places in the region, but fairly widespread on migration and in winter, though not easily seen.

Little Crake *Zapornia parva* 19cm

A very small and secretive crake. Smaller than the commoner Water Rail, similar in plumage, but with a short, green bill which has a red base. Its bill and long wings are important distinctions from the very similar Baillon's Crake *Zapornia pusilla*, not illustrated).

Where to see Although it has bred in the region, it is mainly a rarely seen migrant which also winters in Iraq and southern Arabia.

Spotted Crake *Porzana porzana* 23cm

A small crake, slightly larger than Little Crake with which it can be confused. The heavily white-spotted plumage, buff under the cocked tail and yellow bill with a red base are features to look for. In flight it often dangles its legs.

Where to see A few breed in overgrown wetland vegetation near the Gulf, but otherwise a migrant in spring and autumn with a few remaining in winter.

Grey-headed Swamphen *Porphyrio poliocephalus* 42cm

A large, plump, chicken-sized waterbird with bluish-purple plumage, greyish head and neck and huge red bill, which has a red frontal shield when adult. A shy, slow-moving bird that often seeks cover at the water's edge.

Where to see Resident locally in Syria, the Iraq marshes and along the Gulf coast, where it nests in freshwater swamps and large reedbeds with dense vegetation.

Common Moorhen

Gallinula chloropus 33cm

A dark, hen-like waterbird that is equally happy swimming or walking jerkily along the edge of a well-vegetated wetland. Note the red frontal shield and bill with a yellow tip, white stripe down the sides and white under the cocked tail. Juveniles are brownish.

Where to see Both resident and a widespread migrant, which nests in scattered freshwater wetlands and river margins throughout the region.

juv.

Eurasian Coot

Fulica atra 40cm

Larger than a Common Moorhen and all black with a white bill and frontal shield. Has hunch-backed appearance when it swims. When walking on the edge of a marsh or pool note its green legs and lobed feet. Gathers in large flocks, especially in winter.

Where to see Breeds sporadically throughout the region, but widespread on migration and in winter. Prefers larger areas of water than does Common Moorhen.

juv.

Demoiselle Crane *Grus virgo* 95cm; wingspan 175cm

The smaller and rarer of the two cranes that occur in the region. Distinguished from Common Crane by its black head and neck, which terminates in a fringe of feathers hanging over the breast. Occurs in flocks on migration.

Where to see Mainly on spring and autumn migration in western Arabia and only occasionally elsewhere. Usually in fields close to water and the margins of wetlands.

Common Crane *Grus grus* 115cm; wingspan 235cm

The tallest bird in the region, large and long-necked with a red-topped black head sporting a white stripe behind the eye and running down the black neck. On the ground, note bushy rear end. Occurs in flocks on migration, which in flight are often in a V formation.

Where to see Occurs in fields and at wetlands on migration and in winter in western Arabia, Iraq and Mediterranean countries. Rare elsewhere.

Little Grebe

Tachybaptus ruficollis 27cm

A small diving waterbird with a short neck and fluffy, blunt-ended body. When breeding, note the bright chestnut on the adult's head and conspicuous yellow patch at the base of the bill. Immatures and birds in winter are duller brown. Dives with a fast jump and is reluctant to fly.

Where to see Vegetated lakes and, in winter, estuaries. Breeds in scattered wetlands throughout the region, and is widespread in winter.

juv.

sum.

Great Crested Grebe

Podiceps cristatus 50cm

The largest grebe in the region, with a long neck, pink bill and, in summer, a frill of chestnut and black plumes adorning the head. These are lost in winter, when it shows only a dark grey crest. Dives frequently.

Where to see Breeds in a few isolated colonies in freshwater wetlands in the north of the region south to Qatar. Occurs more widely in winter, including on the sea.

win.

Black-necked Grebe *Podiceps nigricollis* 30cm

A small grebe with a slightly up-tilted bill and steep forehead. The black neck and a spray of yellow behind the eye make it unmistakable in the breeding season. In winter has a clean black-and-white appearance with white cheeks and black cap extending below the ruby-red eye.

Where to see Occurs throughout much of the region in winter on lakes and coastal waters, but there are only a few irregular, isolated breeding colonies.

sum.

win.

sum.

Greater Flamingo *Phoenicopterus roseus* 130cm

Tall and elegant with long pink legs, long curved neck and characteristic salmon-pink bill with a black tip. Adults often acquire a pink hue in breeding season. Young birds are smaller, greyish and with a grey bill. The smaller, Near Threatened Lesser Flamingo (*Phoeniconaias minor*, not illustrated) is all pink with a dark bill and is much rarer.

Where to see A few isolated breeding colonies on coastal lagoons and salt lakes; in winter, more widespread and abundant in some coastal areas.

juv.

Spotted Thick-knee

Burhinus capensis 43cm

Very similar to Eurasian Stone-
curlew in size, coloration, staring
yellow eyes and the way it slowly
walks. But note the black-spotted
and barred plumage (except in
juvenile, which is streaked) and
the absence of a white bar on
the wing.

Where to see Resident in Oman
and southern Arabia in semi-deserts
and low rocky and broken ground,
always with some bushes and trees.

Eurasian Stone-curlew
Burhinus oedicnemus 42cm

An unusual, large, brown and black-streaked ground-dwelling bird with staring yellow eyes, short bill and yellow legs. Walks slowly and in flight note the white patches in its black flight feathers. On migration and in winter often in flocks.

Where to see A breeding summer visitor to countries bordering the Mediterranean, and widespread on passage and winter on steppes, semi-deserts and arable land.

Eurasian Oystercatcher
Haematopus ostralegus 43cm

A large, black-and-white wading bird of coastal shores, which is globally Near Threatened. Note its bright red bill and legs, and when close ruby-red eyes. In flight shows a broad white wing-bar and white lower back. Gathers in flocks and its loud, far-carrying *kleep-kleep* call often indicates its presence.

Where to see A spring and autumn migrant and winter visitor to all coasts and some inland wetlands.

Black-winged Stilt

Himantopus himantopus 37cm

Slim, elegant wading bird with very long, deep-pink legs and black-and-white plumage. In flight note the all-black wings and long legs trailing behind. In breeding season is rather noisy, uttering a high-pitched, repeated *kikikikkik* when disturbed.

Where to see Breeds in scattered colonies in the region on freshwater or saline lagoons, particularly along the Gulf and Red Sea coasts; more widespread on migration and in winter when also found on estuaries.

Pied Avocet

Recurvirostra avosetta 44cm

A delicate, black-and-white wading bird with blue-grey legs and characteristic up-curved bill which it uses to sweep through the surface of a lagoon in search of small crustaceans, its main prey. Noisy when breeding, making a loud *blute-blute* call.

Where to see Breeds in colonies in saline lagoons, mainly near the Gulf. Widespread on migration and in winter on coasts and wetlands throughout the region.

Northern Lapwing

Vanellus vanellus 30cm

Globally Near Threatened and the only wading bird with a long, upturned crest, making it quite distinctive. The white underparts with a broad black breast-band are another important identification feature. In flight note the characteristic rounded wings and floppy wingbeats. Often occurs in flocks.

Where to see A winter visitor to fields, marshes and coasts in northern areas of the Middle East south to the Gulf countries, but rare to Oman.

Spur-winged Lapwing *Vanellus spinosus* 26cm

A rather elegant, long-legged wading bird of wetland margins. Note the black head and underparts contrasting with the white sides of the face and neck. In flight shows a conspicuous black, white and sandy wing pattern. Noisy when nesting.

Where to see Scattered breeding colonies, mostly in the north of the region and along the Red Sea with dispersal in winter, but rarely reaching the Gulf States.

Red-wattled Lapwing *Vanellus indicus* 33cm

Make sure you do not confuse this species with Spur-winged Lapwing. The head pattern of Red-wattled is quite different and note its red bill, yellow legs and white breast and belly. Call is a characteristic, noisy *did he do it, did he do it.*

Where to see Fields, roadside verges and open country, usually near water, in Iraq, the Gulf States and Oman, both in the breeding season and during dispersal in winter.

White-tailed Lapwing *Vanellus leucurus* 26cm

A graceful wading bird with long yellow legs, a rather plain face accentuating the dark eyes, grey breast and all-white tail. In flight the wing pattern resembles that of Spur-winged and Red-wattled Lapwings, which can occur in the same areas.

Where to see Pools and other wet areas, but also cultivation. It breeds, often colonially, in Iraq, Syria and the Gulf States with wide dispersal in winter to Oman and southern Arabia.

Pacific Golden Plover

Pluvialis fulva 24cm

This winter visitor to the Middle East from its Asian breeding areas is rarely seen in summer plumage, which is striking with black face and underparts, bordered by white, separating the black from the golden-spangled upperparts. In winter it is mottled brown and black above and buffish below. Note the short black bill.

Where to see Winters in small numbers on coastal mudflats, plains and cultivation mainly along the Gulf south to Oman.

Grey Plover

Pluvialis squatarola 29cm

In breeding plumage this is a stunning wading bird with its silver-spangled upperparts, black underparts and extensive white on sides of the neck. In winter it becomes mottled grey above and whitish below. In flight, in all plumages, note the diagnostic black patch on the 'arm-pits'. Often in small flocks.

Where to see A passage migrant and winter visitor, mainly to coastal flats throughout the region.

Common Ringed Plover
Charadrius hiaticula 19cm

Small shorebird with black-and-white face markings, black breast-band, white wing-bar and orange legs. In summer the bill is orange with a black tip, but in winter becomes all dark. Often gathers in small groups. See also Little Ringed Plover with which it can be confused.

Where to see A passage migrant and winter visitor throughout the region, favouring sandy and stony shores on coastal flats and inland wetlands.

sum.

win.

sum.

win.

Little Ringed Plover
Charadrius dubius 15cm

Similar to Common Ringed Plover, but smaller with slightly different head pattern and a yellow eye-ring. In flight, note the wings are sandy-brown above, lacking a white wing-bar. Unlike Common Ringed Plover it does not gather in flocks.

Where to see Breeds in scattered colonies in Mediterranean countries, Syria, Iraq and the Gulf south to Oman, preferring freshwater wetlands and dry riverbeds, especially with sand or gravel. Widespread on migration.

Kentish Plover

Charadrius alexandrinus 16cm

A familiar plover of the region's shores. Small, sandy above, white below, with a white neck-ring. Male has a short, black half-collar, which is brown in winter and in the female. Black legs help separation from Common and Little Ringed Plovers.

Where to see Breeds in shingle and sandy areas on the coasts of the Mediterranean, Arabia and the Gulf, and inland in Iraq and Syria. Widespread on migration and in winter.

Lesser Sand Plover *Charadrius mongolus* 20cm

The two sand plovers are very similar and often not easy to identify to species, especially in winter plumage. Lesser is slightly larger than Common Ringed Plover; compared to Greater Sand Plover note shorter bill and shorter, darker legs. In breeding plumage has all-black forehead and extensive reddish breast-band.

Where to see A migrant and winter visitor from Asia to the mudflats and sandy coasts of the Gulf countries, Oman and southern Arabia.

sum.

win.

Greater Sand Plover

Charadrius leschenaultii 24cm

The larger of the two sand plovers, but care must be taken to distinguish between them. Note the larger bill, longer, yellowish-green legs and in breeding plumage the different head pattern.

Where to see Breeds on inland sand- and mudflats in small, scattered colonies mainly in Syria and along the northern Gulf. Otherwise, a widespread migrant and winter visitor to coastal areas of the region.

win.

sum.

Eurasian Dotterel

Charadrius morinellus 21cm

This short-billed, plover-like wading bird is rarely seen near water. In breeding plumage unmistakable with chestnut underparts, bordered by white across the breast – a feature that can be seen in its brownish winter plumage. Also noticeable is the white 'V' on the back of the neck.

Where to see Occurs in winter on steppes and arable fields in northern parts of the region and on the northern coast of the Gulf.

win.

Pheasant-tailed Jacana *Hydrophasianus chirurgus* 31cm (48cm with full tail)

A rare visitor from Asia that is often seen walking on floating vegetation, assisted by its very long toes, which give it the familiar name of 'lily-trotter'. Striking in breeding plumage when note its long, downcurved tail.

Where to see Winter visitor to ponds and marshes with open water and, especially, floating vegetation in Oman (where it has bred) and UAE. Vagrant elsewhere.

win.

br.

Eurasian Whimbrel *Numenius phaeopus* 41cm

A streaked, brown wading bird with a long, downcurved bill. Very similar to Eurasian Curlew, but smaller and with a shorter bill. Note the characteristic stripes on the head and the distinctive seven-note flight call, a fast, whistling *bi-bi-bi-bi-bi-bi-bi*. In flight shows all-brown wings and a white lower back. Often in small flocks.

Where to see Mainly a spring and autumn migrant to estuaries, mudflats and rocky coasts in the whole region.

Eurasian Curlew

Numenius arquata 55cm

Globally Near Threatened, a large, streaked, brown wading bird with a long downcurved bill. Similar to Eurasian Whimbrel but with a much longer bill and no stripes on the head. Far-carrying *cour-leee* flight call is a familiar sound on coastal mudflats. Often in flocks and note all-brown wings and white lower back in flight.

Where to see A common winter visitor and migrant to estuaries, mudflats and rocky coasts in the region; also grassy wetland margins.

Bar-tailed Godwit

Limosa lapponica 38cm

Large, globally Near Threatened wading bird, smaller than Eurasian Curlew and with a long, very slightly upturned bill. Very similar to Black-tailed Godwit, but bill shape different, legs shorter and lacks wing-bars and black tail when seen in flight. Striking in its rusty-red breeding plumage, but most seen will be in their grey-brown winter plumage.

Where to see A winter visitor and migrant to the region's coastal estuaries, mudflats and sandy beaches.

Black-tailed Godwit

Limosa limosa 42cm

This large, globally Near Threatened wading bird is similar to Bar-tailed Godwit, especially in winter plumage. Distinguished by its long, straight bill, longer legs and striking flight pattern with broad white wing-bars and white tail with black terminal band. Often seen in flocks.

Where to see A winter visitor and migrant to estuaries and mudflats around the coasts of the region; also the grassy margins of freshwater wetlands.

win.

A stocky, quick-moving shorebird with a short black bill and orange legs. Has a complicated head pattern of brown, black and white, especially in breeding plumage when its chestnut upperparts are very striking. Watch it constantly turning over vegetation and stones in search of food.

Where to see A migrant and winter visitor to mudflats, sandy shores and rocky coasts in the whole region; also sometimes on grass verges in coastal towns.

win. sum.

Waders

Ruff *Calidris pugnax* 28cm (male); 22cm (female)

Similar size to Common Redshank but more upright stance, smaller head and shorter, slightly drooping bill. Female smaller than male, which has a bright ruff and head plumes of varying colours in breeding plumage. Legs vary from orange to yellowish-green. Rather plain wings and white sides to tail-base in lazy flight.

Where to see A migrant and winter visitor to many of the region's freshwater wetlands and wet grasslands.

♂ win.

late win./spr.

Broad-billed Sandpiper *Calidris falcinellus* 17cm

A small wading bird of mudflats. Slightly larger than the much commoner Little Stint. Similar in shape and size to Dunlin, but with shorter legs and broad-based bill that droops at the tip. To ensure a correct identification it is important to see the yellowish-grey legs and characteristic head stripes.

Where to see Very uncommon or rare migrant and winter visitor to coastal mudflats and estuaries in the region.

Temminck's Stint

Calidris temminckii 13cm

One of the smallest wading birds and very similar to the much commoner Little Stint, from which it is best identified in all plumages by its yellowish-green (not black) legs, defined grey breast and lack of a pale 'V' on its upperparts. Usually found singly or in small flocks.

Where to see A spring and autumn migrant to pools, marshes and coasts in the region, some remaining in winter.

sum.

Curlew Sandpiper

Calidris ferruginea 19cm

This globally Near Threatened wading bird closely resembles Dunlin in size and shape, but note the longer, more decurved bill and longer legs. The grey and white winter plumage resembles Dunlin's, but chestnut summer plumage is quite different. In all plumages shows a white rump in flight.

Where to see A winter visitor and migrant in spring and autumn to estuaries and mudflats around the coasts of the region, especially Arabia.

win.

Sanderling *Calidris alba* 20cm

A small wader, similar to Dunlin in size, which is frequently seen running fast along the shoreline where the waves break. In winter pale grey and white with characteristic dark mark at the bend of the wing. In summer plumage largely a mixture of chestnuts and reds with a prominent breast-band.

Where to see A winter visitor and migrant in spring and autumn to shores, beaches and mudflats around coasts of the region.

Dunlin

Calidris alpina 18cm

A common wading bird in winter when it is one of the most familiar shorebirds on the region's coastal mudflats. In summer plumage the black belly is diagnostic, but in winter the bird is rather drab grey and white with a slightly downcurved black bill and black legs. Often gathers in large flocks.

Where to see A winter visitor and migrant in spring and autumn to estuaries and mudflats around the coasts of the region, especially Arabia.

Waders

Little Stint

Calidris minuta 13cm

One of the smallest and commonest shorebirds of the region, often in large flocks. In winter plumage greyish above and white below with a greyish breast and black legs. In summer plumage it is a mixture of chestnut and buff with a pale 'V' on the back.

Where to see A winter visitor and migrant in spring and autumn to coastal estuaries and mudflats of the region and inland wetlands, but rarer in the north.

Common Snipe *Gallinago gallinago* 26cm

A secretive wading bird with camouflaged plumage and a very long, straight bill which it often probes into the mud with rapid movements. Note the buff and black striped head and buff stripes on the back. When disturbed rises high with fast zigzag flight, uttering a harsh rasping call.

Where to see A migrant and winter visitor to marshes and wetlands throughout the region.

A rather rare wader that may be difficult to spot amongst other shorebirds on the tideline. Look for a small, stocky bird with short orange-yellow legs, a long upturned bill and darting movements, frequently bobbing its rear end as it searches for food.

Where to see Mainly occurs on mudflats and in mangrove creeks around coasts of Arabia on migration and in winter, but only in small numbers.

Red-necked Phalarope

Phalaropus lobatus 18cm

A small, elegant wader more likely to be seen swimming than walking on the shore. It sits high in the water and frequently 'spins' whilst pecking for food that it has stirred to the surface. Distinct, colourful summer plumage, but in winter grey and white with a black eye-mask.

Where to see On migration occurs on the sea and inland waters throughout the region; winters in the Arabian Sea.

sum.

win.

Common Sandpiper

Actitis hypoleucos 20cm

A busy wading bird with a constantly bobbing rear end as it searches for food along a wetland edge. White underparts with characteristic white wedge dividing the brown breast-sides and upperparts. In flight note stiff, shallow wingbeats alternating with short glides.

Where to see Can occur on any wetland, but mostly by fresh water, during spring and autumn migration and in winter when it is rare in the north of the region.

Green Sandpiper
Tringa ochropus 23cm

A rather dark wader (larger than Common Sandpiper), best identified in flight when it shows all-dark wings above and below, these contrasting with the white belly and white base to tail. Note its loud, sharp metallic four-note flight call.

Where to see Can occur on almost any freshwater wetland, stream or small pool throughout the region, on migration and in winter.

Wood Sandpiper *Tringa glareola* 20cm

Similar to Green Sandpiper, but with paler, whitish-speckled upperparts and white stripe above the eyes; also longer, yellower legs which project beyond the tail in flight. Flight pattern similar to Green Sandpiper, but underwings whitish, not dark. When disturbed and put to flight, utters a far-carrying *jiff-iff-iff-iff*....

Where to see Freshwater wetlands throughout the region during spring and autumn migration, often in large flocks, but rare or absent in winter.

Waders

Marsh Sandpiper

Tringa stagnatilis 23cm

A very delicate wader, most closely resembling Common Greenshank but with proportionately longer legs and a fine, straight, needle-like bill. The flight pattern is similar to Greenshank's and care is needed in separating the two.

Where to see A winter visitor and migrant in spring and autumn to the region's coastal estuaries and mudflats and inland wetlands; much rarer in the north, especially in winter.

Common Redshank

Tringa tetanus 28cm

A medium-sized wading bird that is brown above and white below with a brownish breast. One of only two waders with bright red legs, the other being the much rarer Spotted Redshank, from which it is easily distinguished in flight by the broad white trailing edge to the wings.

Where to see Fairly common during migration and in winter on coastal shores and mudflats as well as inland wetlands.

Spotted Redshank
Tringa erythropus 30cm

Taller and slimmer with a longer, finer bill than Common Redshank, the other wading bird in the region with bright red legs. Black, speckled white breeding plumage is distinct, but in winter rather pale grey. Note dark wings in flight with distinctive white wedge on lower back.

Where to see During spring and autumn migration and in winter is found throughout the region, mostly in freshwater wetlands rather than on coastal mudflats. Rather uncommon.

sum.

win.

Common Greenshank *Tringa nebularia* 32cm

A wading bird of shores and mudflats that is larger than Common Redshank, with a long, slightly upturned bill and long green legs. In flight it gives a loud *chew-chew-chew* call, shows all-dark wings and a noticeable white tail extending into a wedge up the back.

Where to see Fairly common during migration and in winter (except in the north) on coastal shores and mudflats as well as inland wetlands.

Crab-plover
Dromas ardeola 39cm

One of the most distinctive birds in the region and one that visitors always want to see. Unmistakable black-and-white wader with blue legs (which extend beyond tail in flight) and a powerful bill that is perfect for crushing crabs, a main prey item. Nests in a tunnel excavated in the sand.

Where to see Breeds colonially in sandy areas around the coasts of Arabia. Disperses to mudflats and coral reefs.

Cream-coloured Courser *Cursorius cursor* 24cm

This desert-dwelling wader is high on the list of species that visitors want to see. Sandy-buff coloration is enhanced by its characteristic back-and-white head pattern, which joins in a 'V' on the nape. It has an upright posture, runs quickly and in flight shows black outer wings and all-black underwings.

Where to see Present in many desert areas in the region throughout the year, either as a breeding bird or passage migrant.

Collared Pratincole *Glareola pratincola* 25cm

An unusual wader that spends much time in the air hunting insects, when it resembles a tern with its long wings, forked tail, short bill and fast, sweeping flight. In flight the chestnut underwings are also distinctive. When perched note the cream-coloured throat.

Where to see Seen mostly on migration, occurring on dried mudflats and grasslands, usually near water; rare in winter. Breeds colonially in a few scattered areas in the region.

Black-headed Gull *Chroicocephalus ridibundus* 38cm

One of the commonest smaller gulls in the region with a dark brown (not black) head in breeding plumage – the only gull to have this. In winter and immature plumage note the dark spot behind the eye, which is only occasionally shown by the similar and rarer Slender-billed Gull. Always has a dark eye.

Where to see Common on all coasts and some inland waters during migration and in winter.

sum.

win.

Slender-billed Gull

Chroicocephalus genei 43cm

Easy to confuse with Black-headed Gull, but never has a black head. Note slightly larger size with longer legs, longer neck and sloping forehead, which accentuates the longer bill. In summer the breast often shows a rosy tinge. In winter usually has greyish spot behind the eye.

Where to see Found on coastal and inland waters during migration and in winter, with a few breeding colonies in the north, especially in Iraq and Kuwait.

Pallas's Gull *Ichthyaetus ichthyaetus* 68cm

Gulls

The largest gull to be seen in the region. Unmistakable in summer plumage with its black head and large yellow bill with red and black markings. In winter loses the black head and then has just a dusky patch behind the eye; in this plumage can be difficult to spot when resting with other large gulls.

win.

Where to see Rather uncommon around the coasts of the region in autumn, winter and spring.

sum.

White-eyed Gull *Ichthyaetus leucophthalmus* 40cm

A Middle East speciality with most of the world's population occurring in the region. Can be easily confused with the similar Sooty Gull, but distinguished by its slender all-dark red bill, black hood and bib and, especially, a noticeable white eye-ring.

Where to see A bird of the Red Sea and rarely seen away from its coasts where it breeds colonially on islands.

juv.

Sooty Gull

Ichthyaetus hemprichii 44cm

Very similar to the rarer White-eyed Gull, but browner, less grey and with a dark brown hood. At all ages the stouter bill is bicoloured – yellowish-green with a dark red and black tip. Often in flocks.

Where to see Resident, breeding colonially on the coasts and islands of the Red Sea, Arabian Sea and the lower half of the Gulf. Often found near ports and fishing villages.

sum.

win.

ellow-legged Gull

arus michahellis 54cm

he large white-headed gulls are
ll very similar and in many cases
will not be possible to make a
onclusive identification. This is
he commonest species on the
Mediterranean coast and in the
xtreme north of the region. Note
he adult's bright yellow legs, yellow
ill with red spot and black ends to
wings with a small white spot on tip.

Where to see Resident on
Mediterranean coast, spreading to
nland lakes in winter. Absent or
ery rare in Arabia.

Armenian Gull

Larus armenicus 56cm

Very similar to Yellow-legged Gull (and
Caspian Gull *Larus cachinnans*, not
illustrated) but adults best distinguished
by more robust, rather blunt-ended bill
with, typically, a broad, black band
near tip. Upperparts are slightly darker
grey, the legs orange-yellow and the
eye dark. The only gull in the region
that is globally Near Threatened.

Where to see A winter visitor to the
Mediterranean coast, northern Gulf
and inland waters in Iraq.

Lesser Black-backed Gull

Larus fuscus 53–65cm

Three races occur regularly in the region. All have black or dark grey backs, distinguishing them from all other gulls likely to be encountered. Adults of the smallest race, *fuscus*, (Baltic Gull) have jet-black upperparts, those of the largest race *heuglini*, (Heuglin's Gull), have dark grey upperparts, whereas those of the race *barabensis* (Steppe Gull, not illustrated) are slightly paler grey.

Where to see Baltic Gull is a winter visitor to all of the region's coasts, whilst Heuglin's and Steppe Gulls are found mainly in the Gulf and Arabian Sea.

heuglini

fuscus

Gull-billed Tern *Gelochelidon nilotica* 38cm

The all-black bill, resembling that of a gull, is the main distinguishing feature of this medium-sized tern which often hawks for insects over mudflats and grasslands. Its black cap, lacking the crest of Sandwich Tern, is reduced in winter to a black eye-mask.

Where to see Migrant on coasts and inland waters throughout the region, with some wintering in southern Arabia. Breeds colonially in north of region, especially in Iraq and the Gulf.

sum.

win.

Caspian Tern *Hydroprogne caspia* 53cm

A very large tern, the size of a gull, with a black crown and large, red bill that immediately distinguishes it. The flight is rather slow and note the dark flight feathers below. Often mixes with other terns.

Where to see There are a few breeding colonies on the coasts and islands of the Red Sea and the Gulf, otherwise found on migration throughout the region, though rare inland.

Greater Crested Tern
Thalasseus bergii 46cm

Slightly larger than (but similar to) Lesser Crested Tern, which occurs in the same coastal regions of the Middle East. Note especially its long, stout, slightly drooping greenish-yellow bill, shaggy black crown and dark grey upperparts. Usually occurs in flocks and plunge-dives for fish.

Where to see Resident colonial breeder on sandy and rocky islands around the coast of Arabia, dispersing widely in winter.

Lesser Crested Tern

Thalasseus bengalensis 41cm

Similar to Greater Crested
Tern, but smaller and lighter in
build and with a slimmer bill
that is distinctly orange-yellow
(not greenish-yellow). The grey
upperparts are slightly paler.
Often occurs in large flocks and
plunge-dives for fish.

Where to see A coastal bird
nesting colonially on sandy or
rocky islands in the Gulf and Red
Sea, dispersing to all coasts of
Arabia in winter.

sum.

win.

Sandwich Tern

Thalasseus sandvicensis 41cm

Similar in size and shape to Lesser Crested Tern, but instantly distinguished by its black bill with a yellow tip and pale grey upperparts with white rump and tail. Often in small flocks mixed with other terns and plunge-dives for fish.

Where to see A migrant and winter visitor to all coasts in the Middle East, with many staying in summer. There are a few colonies on islands in the Gulf.

win.

win.

sum.

sum.

Saunders's Tern

Sternula saundersi 22cm

The smallest tern in the Middle East and almost impossible to distinguish from very similar Little Tern (*Sternula albifrons*, not illustrated). In summer plumage, both have a yellow bill with a black tip and a black cap with a white forehead, but in Saunders's this doesn't extend back to the eye. Also, on Saunders's the outer three wing feathers are noticeably black. Plunge-dives for fish.

Where to see Nests on beaches around the coasts of Arabia. May not be present in winter, when Little Tern is the common species.

Common Tern

Sterna hirundo 35cm

A medium-sized tern that is familiar around the region's coasts. The most important feature for identification is the red bill with a black tip – a feature shown by no other tern in the region. Often in flocks with other terns and plunge-dives for fish.

Where to see Fairly common in spring and autumn around all coasts, but in winter mainly in the south. Nests colonially, mainly in Iraq.

sum.

sum.

White-cheeked Tern

Sterna repressa 33cm

In winter plumage similar to Common Tern when separation is not easy. Note slightly smaller size, more slender bill and grey on upperparts extending onto rump and tail. In summer quite distinctive, being silver-grey above, grey below with a white cheek-stripe.

Where to see A summer visitor nesting in colonies on islands in the Red Sea, the Gulf and off Oman; more widespread on Arabian coasts on migration, but rare in winter.

Whiskered Tern

Chlidonias hybrida 25cm

One of the smaller terns. Distinctive in summer plumage, having grey upperparts and dark grey underparts with a white cheek-stripe below its black cap. In winter confusable with White-winged Tern, but note heavier bill, broader wings, grey rump, streaked crown and shape of cheek-patch.

Where to see Breeding summer visitor to Iraq and Syria, otherwise a passage migrant and winter visitor to coasts and inland waters throughout, but rare in north in winter.

win.

sum.

White-winged Tern *Chlidonias leucopterus* 22cm

One of the smaller terns. Distinctive in summer plumage with its all-black body and white wings and tail. In winter similar to Whiskered Tern, but note slimmer wings, smaller bill, dark spot on cheeks and lighter flight. Usually in flocks.

Where to see Breeding summer visitor to Iraq and on Gulf coast, otherwise a passage migrant and winter visitor to coasts and inland waters throughout, but rare in north in winter.

win.

sum.

Red-billed Tropicbird *Phaethon aethereus* 48cm (plus 50cm tail)

This seabird is rarely seen from land but there is always a chance! You may not easily see the long tail but look out for the red bill (yellow in young birds) and the pattern of black markings on the upperparts.

Where to see Resident breeder on islands and headlands in the Gulf, southern Red Sea and along the Arabian Sea coast, dispersing widely in winter.

Black Stork

Ciconia nigra 95cm

With its long neck and legs, black-and-white plumage and long red bill this large stork is easily identified. Only confusable with Abdim's Stork, which has a white lower back, different-coloured bill and is a bird of southern Arabia. The sight of Black Stork flocks is a migration highlight in Mediterranean countries.

Where to see A migrant in the northern and western Middle East. Very rare in the Gulf countries.

Abdim's Stork

Ciconia abdimii 80cm

Resembles a small Black Stork, but immediately distinguished by the blue bare cheeks, red surround to the eye and short greyish-green (not red) bill. In flight shows a distinctive white lower back. Often in flocks.

Where to see Occurs throughout the year in foothills and plains in southern Arabia, including south Oman; also often at rubbish dumps. Only breeds in coastal Yemen and Saudi Arabia – on trees, rooftops and pylons.

White Stork

Ciconia ciconia 100cm

The most familiar stork and easily identified by its large size, white plumage, black wings and red bill. In flight note the outstretched neck and trailing legs, which are common to all storks. Often seen in flocks, which can be large.

Where to see A few breed in the north of the region (on buildings, trees and pylons); otherwise a migrant to wetlands and farmland throughout with some staying in winter.

Masked Booby
Sula dactylatra 85cm

Superficially similar to Gannet (*Morus bassanus*, not illustrated but familiar over the seas of Western Europe) but in adult plumage has a broad black hindwing, black tail and black mask around the base of the bill. Plunges from a height to catch fish.

Where to see Entirely maritime but can sometimes be seen from the coast of Oman, along the Arabian Sea and into the Red Sea. Breeding colonies are found on rocky islands off Oman and Yemen.

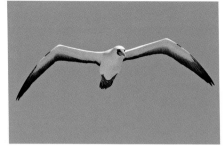

Brown Booby
Sula leucogaster 70cm

Similar to Gannet (*Morus bassanus*, not illustrated) in shape and size, but entirely brown above and with a brown head and neck, contrasting with white body below. Pale greenish-yellow bill is visible at a distance, contrasting with brown head. Flies with series of wingbeats followed by a long glide.

Where to see Throughout the year in the Red Sea, Arabian Gulf and Gulf of Oman. Nests on islands, mainly in the Red Sea.

Pygmy Cormorant

Microcarbo pygmaeus 48cm

A small cormorant with short neck, stubby bill and long tail which is especially noticeable in flight. Plumage glossy black with a bronze sheen on the head. Throat becomes white in winter and young birds are mostly white below.

Where to see Only found in the northern part of the region, wintering on lakes and marshes mainly in Iraq and Syria, and rarely venturing south. Breeds colonially in Iraq and southern Mediterranean countries.

Great Cormorant

Phalacrocorax carbo 90cm

The largest cormorant in the region. In summer plumage has much white on the head and neck and a large white patch on its thighs. In winter only has a white patch on its chin and throat. Can be seen swimming low in the water, frequently diving for fish, or perched with outstretched wings.

Where to see Coasts and lakes throughout the region during migration and in winter.

ocotra Cormorant
halacrocorax nigrogularis 80cm

his Middle East endemic species
classified as globally Vulnerable.
ery similar to Great Cormorant
ut slightly smaller, with a slimmer
ead and neck and lacks white head
narkings. In breeding plumage has
bronzy-green sheen to the back
nd wings.

Where to see A maritime bird that
breeds in dense colonies on islands
n the Gulf and off Oman and
Yemen, dispersing to the coasts and
eas of southern Arabia.

Glossy Ibis
Plegadis falcinellus 65cm

A blackish waterbird with a long,
decurved bill. In breeding plumage
adult has a wonderful purple gloss to
its head, neck and body and a green
sheen to its wings. Winter adults and
juveniles are much duller. In flight
note fast wingbeats interspersed with
long glides.

Where to see Although it breeds
colonially in reedbeds in southern Iraq,
it occurs as a migrant throughout the
region, favouring freshwater wetlands.

Eurasian Spoonbill *Platalea leucorodia* 85cm

The spatulate bill of this large, white waterbird makes it unmistakable. Adult has a black bill with yellow tip, but in young birds it is dull pinkish. Often feeds in small groups, sweeping bill from side to side in the water in search of small fish.

Where to see Shallow fresh and saline water throughout the region on migration and in winter. Breeds colonially on the Red Sea islands, Kuwait's Bubiyan Island and in southern Iraq.

urasian Bittern

otaurus stellaris 75cm

secretive reed-dwelling waterbird.
maller and stockier than Grey
Heron, it is highly camouflaged
n a mix of brown, buff and black
treaks, making it almost invisible
when standing motionless at the
dge of a reedbed.

Where to see Uncommon or rare
n migration and in winter in much
f the region where it is found
n reedbeds and well-vegetated
wetlands. Breeds in southern Iraq.

Little Bittern

Ixobrychus minutus 35cm

A very small, secretive bittern that
is more likely to be seen flying than
creeping through wetland vegetation.
Flight is typically short, on fast
wingbeats, and note especially the
large pale panels on the inner wing
contrasting with black flight feathers
and dark back.

Where to see Well-vegetated ponds,
lakes and rivers throughout the region
on migration and in winter. Breeds
colonially in a few isolated wetlands
and Iraq's marshes.

Black-crowned Night Heron

Nycticorax nycticorax 60cm

Most active at dusk as its name implies. Much smaller and stockier than Grey Heron and unmistakable in its black, white and grey adult plumage. Juveniles are cryptically coloured and most easily identified by their white-spotted brown back and wings.

Where to see Fairly widespread on migration and winter, occurring in well-vegetated wetlands, marshes and rivers. Breeds colonially in a few isolated wetlands in western Arabia, the Iraq marshes and Kuwait.

juv.

Striated Heron

Butorides striata 43cm

A small, dark skulking heron only found on the coast. Often stands crouched and motionless when disturbed. The adult is bluish-grey above with a black crown and greyish below contrasting with rosy or yellow legs. Some colour variation occurs, as shown in the two photos of adults (left). The juvenile is dark brownish, heavily streaked below.

Where to see Resident in mangroves and on rocky and sandy coasts in the southern Gulf and Red Sea with a few breeding on the Arabian Sea coast.

quacco Heron
rdeola ralloides 45cm

small heron, golden-buff in adult
lumage with long black-and-white
lumes on nape. In flight looks
redominantly white as it reveals all-
hite wings and tail. Young birds are
rownish-buff with streaked neck and
reast. Confusable with Indian Pond
Heron (*Ardeola grayii*, uncommon in
UAE and Oman, not illustrated) but
hat species is darker-streaked, has
rown back and lacks buff tones.

Where to see Widespread on migration
n the region's vegetated wetlands,
ivers and ditches, but rarer in winter.
A small number of scattered breeding
colonies.

win.

sum.

Western Cattle Egret
Bubulcus ibis 50cm

You may see this small, active heron
associating with feeding cattle or
other livestock as it feeds on insects
disturbed in grassland. Note the stocky
appearance, short neck, yellow bill
and legs. In the breeding season has
a warm buff wash on head, back and
breast. Usually in small groups.

Where to see Fields and wetlands
throughout the region on migration
and in winter, with scattered breeding
colonies.

Grey Heron *Ardea cinerea* 95cm

A large heron with long neck. Grey upperparts with black crown, white below with black markings down front of neck. In flight note the slow, deep wingbeats, neck tucked into breast and trailing legs. Usually solitary except when breeding.

Where to see Widespread throughout the region on inland and coastal wetlands on migration and in winter. Breeds colonially in trees at a few sites and on Kuwait's Bubiyan Island.

Purple Heron *Ardea purpurea* 80cm

Slightly smaller than Grey Heron with a long, thin, snake-like neck. Much darker in plumage, being a mix of dark grey tones and chestnut when adult and sandy-brown when juvenile. A subtle difference in flight is the more angular neck and spread toes. More secretive than Grey Heron, feeding mostly amongst reeds and other vegetation.

Where to see Marshes and reedbeds throughout the region on migration, with scattered breeding colonies.

he largest all-white heron found in the egion. Very tall, especially when long ngular neck is stretched to full extent. n winter plumage, when most likely to e seen in the region, bill is yellow and egs usually greenish-brown. Mostly solitary. Can be confused with much rarer Intermediate Egret.

Where to see A migrant and winter visitor to coastal and inland wetlands throughout. Breeds in Syria.

Intermediate Egret
Ardea intermedia 65cm

Regularly found only in Oman so it is only there that separation from similar Great Egret is a problem. It can easily be confused but the important features to note are the smaller size, stouter bill and, when seen close, the short gape-line, which does not extend behind the eye. Usually solitary.

Where to see Only found regularly in the coastal and inland wetlands of southern Oman.

Little Egret

Egretta garzetta 60cm

A small, elegant all-white egret found commonly in the region. Easily identified by its all-back bill and black legs with yellow feet. In breeding plumage has long plumes on its nape and breast.

Where to see Throughout the region on migration and in winter on inland and coastal wetlands. Breeds in trees and reedbeds in Iraq and the southern Levant; also on Kuwait's Bubiyan Island.

Hamerkop *Scopus umbretta* 60cm

A strange-looking bird with an apt name (derived from the Afrikaans 'hammer-head'). Its all-brown plumage, strong bill and blunt crest on the nape make it unmistakable. In flight slow wingbeats on rather rounded wings are interspersed with glides. Builds a huge, conspicuous nest of sticks, usually in a tree.

Where to see Only found in Yemen and southern Saudi Arabia where it frequents mountain areas with trees and running water.

Western Reef Heron *Egretta gularis* 60cm

There are two forms (morphs) of this medium-sized heron: a white morph and dark morph. The white morph resembles a Little Egret but note pale yellowish or brownish bill, which is stouter and droops very slightly. The dark morph is unmistakable, being the only slate-grey heron in the region.

Where to see Resident on coasts of the Red Sea, the Gulf and Oman, frequenting tidal wetlands and nesting colonially, especially in mangroves.

dark morph

white morph

Great White Pelican *Pelecanus onocrotalus* 140cm

Pelicans are unmistakable, and this is the commonest of the three species occurring in the region and the most likely to be seen. In flight note all-black flight feathers, distinguishing it from other pelican species. When seen close shows a dark eye surrounded by rosy skin. Often in large flocks, flying in lines or V-formation.

Where to see A migrant to wetlands in the north and west of the region, rare in winter.

Dalmatian Pelican
Pelecanus crispus 150cm

A very uncommon pelican in the region, so take care in identifying it. The most notable features distinguishing it from Great White are greyish appearance, curly nape feathers, pale eye and grey legs; also, in flight, pale band through the centre of the underwing. Pink-backed Pelican (*Pelecanus rufescens*, not illustrated) is rather similar, but only found on Red Sea and Arabian Sea coasts.

Where to see Rare visitor to wetlands in the north of the region.

Western Osprey *Pandion haliaetus* 58cm

This large bird of prey is always found near water, often perched on a tree or post, or dramatically plunging for fish, feet-first. The upperparts are dark brown, the underparts mostly white, whilst the white head has a distinct dark mask through the eye.

Where to see Wetlands, mostly coastal, throughout the region. Breeds on coasts of the Red Sea, the Gulf and Arabian Sea, nesting in trees, cliffs, ruins and even on the ground.

Black-winged Kite *Elanus caeruleus* 33cm

This bird is rare in the region but unmistakable. It is similar to Common Kestrel in size, but its plumage is white below, pale grey above with black shoulders, and wing-ends below. Wings raised when soaring and it frequently hovers.

Where to see Breeds in south-west Arabia and, occasionally, Iraq, otherwise a rare visitor to much of the region, favouring farmland and plains with trees.

Egyptian Vulture *Neophron percnopterus* 62cm

This globally Endangered vulture is easily identified in adult plumage by its white body, white wedge-shaped tail and black flight feathers. When seen close, note the bare yellow face and thin bill. Juveniles mainly dark brownish, but wedge-shaped tail is a clue to identification.

Where to see A migrant through most of the region, often observed at refuse dumps. Breeding populations found in several mountain areas of Arabia, Iraq and countries bordering the Mediterranean.

juv.

European Honey Buzzard

Pernis apivorus 55cm

You often need a close view of this Common Buzzard-sized raptor to attempt identification and for a beginner it is probably best only to try to identify adults. There are three colour forms: pale, dark and intermediate (typical form). All have a rather long tail, small, projecting, pigeon-sized head and characteristic dark banding on the underwing and tail.

Where to see Spring and autumn migrant that can be seen throughout the region but is rare in eastern Arabia.

Griffon Vulture

Gyps fulvus 100cm

One of the largest birds of prey in the region. Groups can be seen soaring for long periods, when note the long, broad, deeply fingered wings held in a shallow 'V' and short protruding head. Gingery-buff plumage is relieved by black flight feathers. Gathers at carcasses and refuse dumps to feed.

Where to see Although it occurs throughout the region on migration, it is rare away from its breeding areas in the mountains of western Arabia, Israel, Jordan and Iraq.

Lappet-faced Vulture

Torgos tracheliotos 105cm

Globally Endangered, this large, blackish vulture is one of the region's most threatened birds. Unlike Griffon Vulture, usually seen singly or in pairs. When perched note heavy bill, bare pinkish head and tufts of feathers on hindneck and breast. Similar Cinereous Vulture (*Aegypius monachus*, not illustrated) is all black with whitish head.

Where to see Mostly resident in the savannas and semi-deserts of central Arabia with populations also in UAE, Oman and Yemen. Rare or vagrant elsewhere.

Short-toed Snake Eagle

Circaetus gallicus 68cm

A large, long-winged eagle, very pale below and typically with a dark brown head and upper breast. Frequently shows light barring on the underwing and always has bars on the tail. Often hovers in search of ground-dwelling prey, especially snakes and lizards.

Where to see Occurring on migration throughout the region, this mostly summer visitor breeds at scattered localities in Arabia, northern Iraq and countries bordering the Mediterranean.

Lesser Spotted Eagle

Clanga pomarina 62cm

Large, medium-brown eagle, very similar to Greater Spotted and adult Steppe Eagles, and separation can be very tricky. Lesser Spotted is the smallest of the three and note its paler brown upperwing-coverts contrasting with darker back. Migrates in large flocks which circle to gain height in the thermals over mountain slopes.

Where to see Spring and autumn migrant commonly seen only when migrating through the mountains of countries bordering the Mediterranean.

Booted Eagle *Hieraaetus pennatus* 48cm

A small eagle, similar in size to Common Buzzard, and confusingly with two types: pale morph and dark morph. The pale morph is white below with black flight feathers, whilst the dark morph is dark brown with pale wedges on the underwing.

Both morphs have diagnostic white 'headlights' at the base of the wings.

Where to see A migrant in spring and autumn throughout the region, though often uncommon. Very few overwinter.

Greater Spotted Eagle
Aquila clanga 64cm

This globally threatened eagle is very similar to Lesser Spotted Eagle. Juvenile birds are more easily distinguished by the characteristic rows of white spots on the dark brown upperwing. Adults differ from Lesser Spotted by the all-dark brown plumage with the hindwing below being a shade paler (reverse in Lesser Spotted).

Where to see A rather uncommon migrant throughout the region; winters in small numbers when mostly seen around wetlands, mountains and refuse tips.

Steppe Eagle

Aquila nipalensis 75cm

Globally Endangered and one of the largest eagles to occur in the region. The adult is dark brown, like a large Lesser Spotted Eagle, but when perched note the heavy bill and long, yellow gape-line extending to rear of eye. Juveniles distinctive, with a broad, white band on underwing.

Where to see Migrant and winter visitor throughout the region, most often seen in winter at marshes and refuse tips, where it can gather in large numbers.

juv.

astern Imperial Eagle

quila heliaca 75cm

very large eagle, classified as
obally Vulnerable. In dark brown
dult plumage note the buff hind-
eck, white 'braces' and pale base
the tail. Juvenile birds, which
ost commonly occur in the region,
re yellowish-buff below with a
ark-streaked upper breast and dark
indwings which show a distinct pale
/edge in the outerwing.

Where to see A migrant and winter
isitor throughout the region, but rare.
avours marshes, mountains, wooded
esert and refuse tips.

juv.

Golden Eagle

Aquila chrysaetos 78cm

Large eagle with powerful flight and,
unlike other large eagles, soars and
glides on raised wings. The dark
brown adult has a golden-buff hind-
neck and pale panel on upperwing.
The juvenile is distinctive, with large
white wing patches and a white tail
with a black band at tip.

Where to see Resident in widely
scattered areas in the region, notably
Arabia including Oman and UAE,
and the southern Levant. Breeds
in mountains and semi-deserts
with trees.

Verreaux's Eagle

Aquila verreauxii 90cm

Distinctive large eagle with a unique wing shape: broad, bulging wings pinched-in at the body. The adult is immediately identified by its black plumage with white back and 'braces', and white wing patches above and below. The brownish juveniles have pale wing patches and a blackish-brown throat and breast contrasting with the buffish-white rear body.

Where to see Resident, but rare, in the desolate mountains of western Arabia, Yemen and southern Oman.

Bonelli's Eagle *Aquila fasciata* 65cm

Medium-sized eagle, larger than Common Buzzard, with longer tail and wings that are held flat when soaring and gliding. Adults identified by dark underwings contrasting with whitish body and leading edge of wings, and also by white patch on back. Juveniles have a pale rusty underbody and wing-coverts, bordered by a dark bar. Always occurs singly or in pairs.

Where to see Resident in widely scattered rocky mountains and wooded hills in the region.

Levant Sparrowhawk
Accipiter brevipes 35cm

Similar in its flight actions to the more familiar and widespread Eurasian Sparrowhawk but has longer, more pointed wings. Distinctive male is white below with pale orange barring on breast and black wing-tips. Female similar, but stronger barring below and less distinctly marked.

Where to see Best seen in the countries bordering the Mediterranean, through which soaring flocks pass on spring and autumn migration following the line of hills. Rare elsewhere in the region.

♂

Eurasian Sparrowhawk *Accipiter nisus* 35cm

Unlike Levant Sparrowhawk it never gathers in flocks. A small hawk with short, rounded wings and long tail. In flight rapid wingbeats are interspersed with short glides. The male is slate-grey above with rufous barring below. Female more brownish-grey above with brown barring below. Lacks black wing-tips of Levant Sparrowhawk.

Where to see Widespread migrant and winter visitor to the region, occurring anywhere with trees.

juv.

♀

Western Marsh Harrier

Circus aeruginosus 52cm

The largest harrier in the region. Note wavering flight with raised wings, low over the ground, typical of all harriers when hunting. The male is tricoloured: brown upperparts and forewing, grey wings and large black wing-tips. Female dark brown with cream crown, throat and shoulders.

Where to see Can occur at any marshes, reedbeds and agricultural land throughout the region on migration and in winter.

Montagu's Harrier

Circus pygargus 45cm

Similar to Pallid Harrier, but male distinguished by darker grey upperparts with more extensive black on wing-tips, a black band on upperwing and two bands below. Note also rufous streaking below. Experience is needed to tell females and juveniles (so-called 'ring-tail' harriers) from Pallid Harrier, so most will remain unidentified.

Where to see Wetlands, steppes and agricultural land throughout the region on migration. A few winter in southern Arabia.

allid Harrier
ircus macrourus 44cm

Near Threatened species similar
n shape to Montagu's Harrier. Male
istinguished by pale grey upperparts,
white underparts and black wedge on
ving-tips. Female and juvenile very
imilar to Montagu's, being brownish
bove with white band at tail-base,
ightly streaked below with black
ands on underwing flight feathers.
Only juveniles easily identified by
heir broad, pale collar.

Where to see Wetlands, steppes
nd agricultural land throughout the
egion on migration and in winter,
hough uncommon.

♂

♀/imm.

♀/imm.

Black Kite

Milvus migrans 58cm

Apart from the extremely rare Red Kite, the only raptor with a forked tail found in the region. Plumage dark brown with slightly paler areas near end of underwing. Yellow-billed Kite (*Milvus aegyptius*, not illustrated) is paler with a yellow bill and only occurs in south-west Arabia.

Where to see Common on migration, either passing in flocks, especially through the countries bordering the Mediterranean and Red Sea, or in gatherings at refuse dumps.

Long-legged Buzzard

Buteo rufinus 63cm

Larger than Common Buzzard, at times resembling an eagle in shape. Soars on raised wings. Has three colour forms, but the most typical has a pale head and breast, dark belly and pale, orangey tail – the latter being a key identification feature.

Where to see A breeding resident in scattered areas throughout the region, in plains, semi-deserts and mountains with trees. More widespread on migration and in winter.

Common Buzzard

Buteo buteo 50cm

The race in the region is the highly migratory Steppe Buzzard (*Buteo buteo vulpinus*, not illustrated). Occurs in many colour forms, the most typical having brownish forewings below, with the rest of the wings whitish with a dark trailing edge. Unlike Long-legged Buzzard the darker rufous tail usually has a dark terminal band.

Where to see Common on migration, especially when soaring flocks pass through the hills of countries bordering the Mediterranean and Red Seas. Less common in southern Arabia.

Western Barn Owl *Tyto alba* 35cm

A very pale owl both when perched and in its soft, wavering flight. Note lightly-spotted white underparts and heart-shaped face with prominent black eyes. Mostly nocturnal, but often seen hunting at dusk.

Where to see A breeding resident that is patchily distributed throughout the region, dispersing widely in winter, but nowhere common. Occurs in farmland, semi-deserts, woodland edges and often near buildings.

Eurasian Scops Owl

Otus scops 20cm

A small owl, the most widespread of the region's three scops owls: Eurasian, Pallid (*Otus brucei*, not illustrated) and Arabian (*Otus pamelae*, not illustrated). All are nocturnal and almost identical except for their songs. Eurasian Scops repeats *pwoo*; Pallid has a dove-like, repeated *whoop*; and Arabian repeats *da-pwoorp*.

Where to see Migrant and summer visitor to the region, breeding in wooded areas, including olive groves and oases in countries bordering the Mediterranean. (Pallid Scops breeds in Oman and UAE; Arabian in south-west Arabia.)

Pharaoh Eagle-Owl

Bubo ascalaphus 50cm

Smaller than the browner Eurasian Eagle-Owl (*Bubo bubo*, not illustrated) and largely replaces it in Arabia. Distinguished by prominent ear-tufts, whitish distinctly barred underparts and orange eyes in a pale face. A loud, booming *boooor* call at night.

Where to see Resident in mountains, on cliffs and in deserts with a patchy distribution throughout much of Arabia, except the south-west.

Arabian Eagle-Owl
Bubo milesi 45cm

A large owl with ear-tufts that is similar in shape to Pharaoh Eagle-Owl, but immediately identified by densely barred dark brown underparts and yellow eyes. The upperparts are also dark brown and note fine white spotting on the hind-neck. Often seen in the day, but at night male detected by voice: a repeated two-note *hu-hoo*, often in duet with female.

Where to see Resident in south-west Arabia and parts of Oman, in woodlands and rocky hills.

Desert Owl *Strix hadorami* 37cm

This is the owl of the region's deserts and arid places. It resembles a small, very pale Tawny Owl (*Strix aluco*, not illustrated, familiar in Europe) with orange eyes set in a whitish face. Underparts are also whitish with pale orange-buff barring. Listen out for characteristic five-syllable song: *whooo, hoo-hoo, hoo-hoo.*

Where to see Patchily distributed throughout much of Arabia as well as southern Mediterranean countries; found in arid rocky mountains and deserts, also palm groves.

Little Owl

Athene noctua 22cm

A small owl with rounded head and flat crown that can often be seen in the day as it perches prominently on a post or mound, often 'bobbing' up and down. Plumage can vary from brown through to pale sandy-grey, the latter typical of birds in desert areas. All have boldly streaked underparts and white spotting on head and nape. Note undulating flight.

Where to see Patchily distributed throughout the region, occurring in a wide range of habitats.

Short-eared Owl

Asio flammeus 38cm

Medium-sized owl that often hunts in the day, flying on slow, high wingbeats with spells of gliding on raised wings. Often sits on the ground when note staring yellow eyes. Ear-tufts can be difficult to see, unlike Long-eared Owl (*Asio otus*, not illustrated and mostly a winter visitor to Mediterranean countries) which has long upright tufts.

Where to see Migrant and winter visitor throughout much of region, though rare in southern Arabia. Occurs in open country, especially marshes, including coastal areas.

urasian Hoopoe *Upupa epops* 28cm

)ne of the most distinctive and familiar irds of the Middle East. The pinkish-uff plumage with a black-tipped crest nd black-and-white striped wings and ail make it immediately obvious. Note s broad, rounded wings in flight.

Where to see Breeds in several parts of the region, but most notably in the countries bordering the Mediterranean and Red Seas. Otherwise, a common migrant and winter visitor.

Indian Roller *Coracias benghalensis* 30cm

Similar to European Roller and both can occur in the same area. Most readily identified in flight by striking wing pattern of pale blues divided by dark blue centre to wing. In its breeding area makes spectacular twisting aerial manoeuvres.

Where to see Breeds and is largely resident in southern Iraq, eastern UAE and northern Oman where it is found in open country, parks and gardens with trees. Disperses along the Gulf coast in winter.

European Roller *Coracias garrulus* 30cm

The widespread roller in the region on migration. The colour pattern of turquoise-blue with a chestnut back is common to the three rollers in the region, but note the blue forewing in flight contrasting with dark hindwing (compare with Indian Roller) and no tail streamers (as in Abyssinian Roller *Coracias abyssinicus*, not illustrated, of southern Arabia).

Where to see Breeds in northern Iraq and the Levant, otherwise a migrant to the region, found in open country.

Grey-headed Kingfisher *Halcyon leucocephala* 20cm

A fairly large kingfisher, often found in areas without water. The greyish head and large red bill make it distinctive, as does the large white wing-panel when seen in its rather slow, slightly undulating flight. Often seen on a prominent perch searching for prey, mainly lizards and large insects.

Where to see Mainly a summer visitor to wooded wadis in the highland slopes of southern Arabia, including south Oman.

White-throated Kingfisher

Halcyon smyrnensis 26cm

A large kingfisher, easily identified by its shining turquoise and brown plumage, white bib and very large red bill. Sits prominently looking for prey on the ground. Listen for its loud, raucous call: *kril-kril-kril-kril*.

Where to see A resident breeding bird in southern Mediterranean countries, Iraq and Kuwait, with some dispersal in winter to nearby countries. Occurs on lakes and rivers with trees and also palm groves and dry woodlands.

ollared Kingfisher

diramphus chloris 24cm

nother large kingfisher, but which is
und only in mangroves. Turquoise
ove, white below with a white
llar (hence its name). Mostly
eds on crabs. Its presence often
itially becomes obvious by its
ud, distinctive call: a series of fast,
scending ringing notes.

Vhere to see A breeding resident in
angroves at Khawr Kalba in UAE,
ree isolated coastal areas in
man and the very southern Saudi
rabian coast.

Hoopoe, rollers and kingfishers

Common Kingfisher *Alcedo atthis* 17cm

A small kingfisher identified by its shining blue and green upperparts, reddish-chestnut underparts and blackish bill. Unlike the other kingfishers in the region, it is not a conspicuous bird, often perching quietly on a branch for long periods waiting to dive for fish.

Where to see A winter visitor to coastal and inland wetlands throughout northern and western parts of the region, south to Oman.

Pied Kingfisher

Ceryle rudis 25cm

The largest kingfisher in the region and the only one that is black and white, so quite unmistakable. Listen for its loud calls: a chattering *chirrik*, *chirrik*, *chirrik*. Often occurs in small groups, it hovers above the water before diving headlong for fish.

Where to see Breeds in countries bordering the Mediterranean, also Iraq, dispersing in winter especially along the Gulf. Found on rivers, lakes and on the coast.

rabian Green Bee-eater *Merops cyanophrys* 24cm

e smallest bee-eater in the region
d easily identified by its green
umage with a blue throat bordered
y a black eye-stripe and necklace
elow. In flight reveals pale chestnut
nderwings. Usually in pairs. Perches
w above ground from where it makes
erial sallies in search of insects.

Where to see Resident in farmland,
semi-deserts and parks from southern
Mediterranean countries and much of
western Arabia to Yemen and south
Oman; also north Oman and UAE.

Blue-cheeked Bee-eater *Merops persicus* 30cm

A green bee-eater with long central tail feathers, blue cheeks and a chestnut throat. When it flies reveals chestnut-orange underwings. Often found in large groups, especially on migration.

Where to see Occurs on migration throughout the region; also a colonial breeding summer visitor to Iraq and a few isolated areas in southern Mediterranean countries, Oman and UAE. A bird of dry open country with trees, nesting in holes in sandy ground.

European Bee-eater

Merops apiaster 27cm

Distinguished from all other bee-eaters by its chestnut crown and back, yellow throat and turquoise underparts. Travels in flocks on migration, constantly calling – a soft, far-carrying *prroop*.

Where to see Can occur anywhere on migration in the region; also a colonial-breeding summer visitor to Iraq and a few isolated sites in Mediterranean countries and Oman. A bird of open country with trees, nesting in holes in sandy banks.

Eurasian Wryneck *Jynx torquilla* 16cm

A rather strange bird which is closely related to the woodpeckers. It has a complicated, vermiculated pattern, but note the black eye-stripe, black stripe from the crown broadening down centre of back, yellowish-buff throat and finely barred underparts.

Often feeds on the ground and can be very secretive.

Where to see Can be found throughout the region in any habitat on spring and autumn migration, though often not easily seen.

Arabian Woodpecker

Dendrocopos dorae 18cm

This Near Threatened species is the only true woodpecker found in Arabia. Rather small, it is easily overlooked as it feeds amongst the branches of *Acacia* trees. The olive-brown plumage contrasts with black-and-white banded wings revealed in its undulating flight. Note the male's red belly patch and red crown and nape.

Where to see Only found in the mountains and foothills of western Saudi Arabia and Yemen, where it is resident in woodland, especially *Acacia*.

♂

Syrian Woodpecker

Dendrocopos syriacus 23cm

One of only two black-and-white woodpeckers in the region and is much the commoner and more widespread. Adults distinguished from Middle Spotted Woodpecker (*Dendrocopos medius*, not illustrated, which only occurs in Syria) by the lack of streaking below and black crown (red in Middle Spotted).

Where to see Resident in woodland, parks and gardens with trees in countries bordering the Mediterranean. (Middle Spotted is a bird of wooded coastal hills of northern Syria.)

♂

esser Kestrel

alco naumanni 30cm

esser Kestrel is a small falcon, similar
) Common Kestrel. Only the males
re easily identified. Lesser Kestrel
as an unspotted chestnut back and
orewings, the latter bordered by a
ale grey panel. When perched note
ack of a moustachial stripe. Takes
nsects in flight and is gregarious at
reeding sites.

Where to see Spring and autumn
migrant with breeding colonies in the
countries bordering the Mediterranean,
northern Syria and northern Iraq.

♀

♂

Common Kestrel

Falco tinnunculus 35cm

Very similar to Lesser Kestrel and only males are easily separated. Common Kestrel is distinguished by black spotting on its chestnut back and forewings, more heavily marked underwings and moustachial stripes. Frequently hovers in search of prey.

Where to see A widespread resident in any open habitat in many parts of the region, but otherwise can be seen throughout on migration and in winter.

Red-footed Falcon

Falco vespertinus 30cm

This small, Near Threatened falcon is found only in the Mediterranean region when on migration. The slate-grey male with red thighs and undertail-coverts is readily identified. In the female note the rusty-yellow underparts and head with a black eye-mask. Often found in groups and hunts flying insects, which it grasps in its talons.

Where to see A migrant in spring and autumn through the countries bordering the Mediterranean, but not commonly observed.

ooty Falcon *Falco concolor* 34cm

his medium-sized, long-winged falcon
s considered Vulnerable globally.
Adults identified by their slate-grey
upperparts and blue-grey underparts.
uveniles are rusty below, lightly
treaked on breast and with a dark
moustache. Breeds in late summer and
catches migrating birds to feed young.

Where to see A summer visitor,
breeding on islands, cliffs (including
some inland desert cliffs) in southern
Mediterranean countries, along the Red
Sea, the Gulf and in northern Oman.

juv.

Eurasian Hobby
Falco subbuteo 33cm

A small, fast-flying falcon with scythe-like wings. Upperparts are slate-grey and underparts whitish, heavily streaked and with red thighs and undertail-coverts. Note the black moustache on noticeably white cheeks. Agile when hunting small birds and insects in flight.

Where to see A widespread spring and autumn migrant throughout the region with a few breeding in scattered woodland in countries bordering the Mediterranean.

Peregrine Falcon *Falco peregrinus* 38cm

Large, powerful falcon and the fastest bird in the world, with stoops to catch prey of 200km/hour. Adult has a black crown and obvious black moustaches, contrasting with white cheeks. Upperparts dark blue-grey and underparts white with fine barring on breast and belly. Juvenile is brown with streaked underparts.

Where to see A passage migrant and winter visitor to many parts of the region, but nowhere common.

Rose-ringed Parakeet *Psittacula krameri* 42cm

This parakeet is not native to the region but is now fairly widespread as the result of escapes from captivity. It is easily identified by its green plumage, long tail and, in the male, a rosy ring edged with black around its neck. Gregarious and very noisy, flocks uttering a loud, rather piercing *kee-ak*.

Where to see Occurs in scattered colonies throughout the region, favouring gardens and open woodland, nesting in a hole in a tree.

Black-crowned Tchagra

Tchagra senegalus 22cm

A very secretive, shrike-like bird that is more often heard than seen as it utters its harsh *shrrr* alarm call or fluty and melodious song from deep cover – or, if you are lucky, during a songflight. If seen well, note the black-and-white head pattern, chestnut wing patches and long, graduated white-tipped tail.

Where to see A resident of dry scrub in coastal areas of south-west Arabia and southern Oman.

Red-backed Shrike

Lanius collurio 18cm

♂

The male is a handsome bird with a grey head and nape, chestnut upperparts and striking black eye-mask. Female is browner with a more subdued pattern. Perches prominently on a tree or post from which it drops on prey.

Where to see A widespread spring and autumn migrant throughout the region, favouring open, scrubby habitats with scattered trees.
Only found breeding at scattered localities in countries bordering the Mediterranean.

Isabelline Shrike

Lanius isabellinus 17cm

Resembles Red-backed Shrike, but with pale sandy-grey upperparts and an orange-red tail and rump, thus similar to Red-tailed Shrike; it may not always be possible to separate the two species, especially females and immatures.

Where to see A migrant and winter visitor to much of the region, but nowhere common and rare in the west; found in open country with trees, scrub and parks.

...ed-tailed Shrike

...nius phoenicuroides 17cm

...ery similar to Isabelline Shrike and ...can be very difficult to separate ...e two species. In spring note the ...ifous crown, white supercilium, ...arker upperparts and black eye-mask ...xtending in front of the eye to ...e bill.

...Vhere to see A spring and autumn ...nigrant that can be found in ...teppes, semi-deserts and fringes of ...ultivation, especially favouring areas ...vith acacia woodland. Rarely seen ...n winter.

Lesser Grey Shrike

Lanius minor 20cm

Smaller than Great Grey Shrike, but similar in its grey-and-black plumage. In the adult main distinguishing features from Great Grey are the extensive black over the forehead and pinkish wash to the underparts. Also, Lesser Grey has a stouter bill and broader white wing-panel.

Where to see A migrant in spring and autumn throughout the region; found in open country, cultivation and steppe with scattered trees.

Great Grey Shrike

Lanius excubitor 25cm

A grey, black and white shrike, which is larger than the otherwise similar Lesser Grey Shrike. Adults can be distinguished by the black eye-mask not extending over the forehead, white underparts and longer tail. Often spends long periods on a prominent perch.

Where to see A resident in much of the region, but also a widespread migrant and winter visitor, favouring open country with trees and scrubby areas, particularly with acacias.

Woodchat Shrike

Lanius senator 18cm

Adults are easily identified by their chestnut crown and hind-neck, and black upperparts with large white shoulder-patches. Immature plumage in autumn can look similar to young Red-backed Shrike but note pale wing-patch in flight. Distinguished from young Masked Shrike by finely scalloped warm grey-brown (not cold grey) upperparts.

Where to see Breeds in countries bordering the Mediterranean and in northern Iraq, typically in open country with trees and olive groves. Widespread on migration.

Masked Shrike

Lanius nubicus 18cm

The handsome male is easily identified by its black-and-white upperparts, white face with black eye-mask and white underparts with an orange wash on the sides of the breast. Female similar, but duller. Young birds resemble young Woodchat Shrike but with greyer upperparts.

Where to see Breeds in countries bordering the Mediterranean and in northern Iraq, in open woodland, olive groves and open country with trees. Widespread on migration.

Eurasian Golden Oriole

Oriolus oriolus 24cm

The male, with its bright yellow-and-black plumage, is unmistakable but can be hard to see if it is perched in a leafy tree. Female and immature birds are greenish with brown wings and lightly streaked underparts. On breeding grounds, more often heard than seen, its loud, fluty three-note yodel revealing its presence.

Where to see Breeds in woodland in countries bordering the Mediterranean, northern Iraq and isolated areas in northern Arabia. Widespread on migration.

African Paradise Flycatcher

African Paradise Flycatcher

Terpsiphone viridis male 35cm (including long tail); female 17cm

With its chestnut upperparts, glossy blue-black head and breast and, when full-grown, very long white tail, the male is unmistakable; it also has large, white wing patches. The female is similar but lacks the long tail-streamers. There is a very rare white morph. Slow, rather heavy, wavering flight.

Where to see A breeding resident that is found only in the semi-tropical mountain woodlands of south-west Arabia and southern Oman.

♀

♂

Eurasian Jay

Garrulus glandarius 33cm

A distinctive bird with its pinkish-brown plumage, black crown and face markings, white rump and shining blue wing patches in its black-and-white wings. When breeding can be very shy, but noisy, rasping screeching calls draw attention to it. Note broad wings in flight.

Where to see Largely resident in woodlands, particularly conifers and oak, in countries bordering the Mediterranean and in northern Iraq. Some dispersal in winter when birds also gather in flocks.

Eurasian Magpie *Pica pica* 48cm

A glossy black-and-white bird with a long tail and noisy, chattering calls. Cannot be mistaken for any other species. The very similar Arabian (Asir) Magpie (*Pica asirensis*, not illustrated) is globally Endangered.

Where to see Breeds in wooded mountains in northern Iraq with some winter dispersal to the south. The Arabian (Asir) Magpie is found only in one small area of southern Saudi Arabia.

Western Jackdaw

Coloeus monedula 33cm

The smallest of the crows and identified by grey on the nape extending in a collar to the neck-sides and, when close, white eyes (blue in immature birds). Often in flocks with other crows and on the ground moves quickly when searching for food.

Where to see Breeds in countries bordering the Mediterranean, northern Syria and northern Iraq, often found near habitation as well as farmland and cliffs. Disperses south in winter.

House Crow

Corvus splendens 43cm

Not a native crow, having been introduced by travelling aboard ships. It spread rapidly and has caused serious ecological problems in many areas, in some cases impacting on native wildlife. Note the domed crown, black plumage with grey neck and breast demarcated from black face. Gregarious and very noisy.

Where to see Resident in coastal areas along the Red Sea, Arabian Sea and Gulf, where it is found in ports, towns and villages.

ooded Crow

orvus cornix 47cm

is crow is easily identified by
; grey-and-black plumage –
te the black head and breast
early demarcated from the
ey body. Often found in flocks
utside the breeding season and
ten associates with livestock.
las a croaking *kraar, kraar* call.

Where to see Breeds only in
ie countries bordering the
Mediterranean and in northern
yria and Iraq, mainly in
armland, light woodland
nd hills.

Brown-necked Raven *Corvus ruficollis* 50cm

A large member of the crow family
which is especially associated with
desert areas. Slightly smaller than
Northern Raven (*Corvus corax*, not
illustrated, which occurs in mountains
in the north of the region) and can be
distinguished by the bronzy-brown
sheen on its nape and neck, though
this can sometimes be difficult to see.

Where to see Resident throughout
much of Arabia in deserts, semi-deserts
and dry mountains, often near human
habitation.

Fan-tailed Raven *Corvus rhipidurus* 47cm

This small, all-black raven is best identified in flight. Its very short tail and the bulging rear edges to its wings make its silhouette unmistakable. It is frequently found in very large flocks and will soar on thermals for long periods.

Where to see Found in a range of habitats from sea level to mountain tops, but often close to human settlements, in western, southern and central Arabia, and in south Oman where it replaces Brown-necked Raven.

Grey Hypocolius
Hypocolius ampelinus 23cm

One of the region's specialities, much sought after by birdwatchers. The male is distinctive with its grey plumage, long black-tipped tail, wide black eye-mask and white-tipped wings – prominent in flight. The sandy-grey female lacks the eye-mask.

Where to see Breeds in southern Iraq and Kuwait, in sparse woodland (especially tamarisk) near to water. In winter migrates south to eastern Arabia and Oman, frequenting scrub, date palms and especially berry- and fruit-bearing trees.

ombre Tit
ecile lugubris 14cm

milar in size and behaviour to the
ore familiar Great Tit, but greyish
bove and off-white below with a
lack cap and larger black bib. It
as a distinctive chattering call note
hurr-er-rrrr, which often first attracts
ttention to the bird.

Where to see Occurs only
n countries bordering the
Mediterranean and in northern
aq where it frequents a variety of
woodland and scrub habitats from
lains to mountain slopes.

Great Tit
Parus major 14cm

This bird is well known across
much of Europe and Asia but occurs
only in the north of our region. It is
very easily identified by its yellow
underparts with a black stripe down
to the belly and its black head with
large white cheeks.

Where to see Occurs only
in countries bordering the
Mediterranean and in northern Iraq
where it is found in woodlands and
gardens and can be quite tame.

Eurasian Penduline Tit *Remiz pendulinus* 11cm

Nearly always found near water and identified by the male's chestnut back, a black mask through the eyes and a fine bill adept at taking seeds from reeds and bulrushes. It builds an oval-shaped nest suspended over water from an outer branch of a tree.

Where to see Only found breeding in a few sites near the Mediterranean coast and Syria, in wetlands, including rivers, with trees, especially willows and tamarisk. Disperses more widely in winter.

Desert Lark

Ammomanes deserti 15cm

Rather nondescript, unstreaked, sandy-grey lark with broad wings and a short tail. Its bill is fairly stout. In rather floppy, undulating flight note the dark brown tail with rufous base. Much variation in colour with some races sooty-grey (notably the race *annae* in the Jordan basalt desert) while paler ones are pinkish-buff.

Where to see Resident mostly found in dry areas, especially rocky or stony slopes with little vegetation, throughout much of the region.

One of the largest larks in the region and found only in sandy areas. Note its upright stance, long decurved bill and, when it flies, its startling white-and-black wing pattern. It runs swiftly, making sudden stops, and has a remarkable rising and twisting song-flight, steeply diving before landing on outstretched wings. Its melodious and melancholy song is a delight of dawn in the desert.

Where to see Widespread resident throughout much of the region in deserts, semi-deserts and dunes.

Bar-tailed Lark *Ammomanes cincturus* 13cm

Smaller than Desert Lark, but very similar in plumage and the two species can often be confused. The most important difference is the tail pattern, which is best seen in flight when note the clear-cut black band on a reddish-brown tail.

Where to see Patchily distributed in desert, semi-desert and stony areas throughout much of the region, but absent from much of Iraq and southern Arabia, though present in central Oman and UAE.

Black-crowned Sparrow-Lark *Eremopterix nigriceps* 12cm

The male is unmistakable with all-black underparts and white cheeks. The female is rather featureless with unstreaked sandy-grey upperparts and buffish underparts, but note the stout pale grey bill. Often found in flocks feeding on bare ground.

Where to see Sandy plains and semi-deserts with scattered vegetation, also cultivated fields. It is found mainly through the coastal areas of Arabia. Absent from the north of the region.

Singing Bush Lark
Mirafra cantillans 15cm

Most easily identified in its low, weak, fluttering flight when note the broad wings with rufous flight feathers and short tail. In early morning best noticed by its bat-like song flight. Otherwise, a rather featureless, streaked lark with fairly stout, yellowish bill.

Where to see Only found in the extreme south of the region, notably on the coastal plains, farmlands and semi-deserts of Yemen and southern Oman, where it is a common resident.

Woodlark *Lullula arborea* 15cm

Small, very short-tailed lark with noticeable whitish supercilia that join on the back of the head in a V-shape and black-and-white marking on each wing. The short tail is especially noticeable in its deeply undulating flight.

Where to see Only found in the north of the region where it breeds in countries bordering the Mediterranean, also northern Iraq and Syria, favouring open woodlands. Disperses south in winter when it frequents fields, but rarely reaches Arabia.

Eurasian Skylark

Alauda arvensis 18cm

Grey-brown streaked lark, similar to Crested Lark in size and plumage but only has a very short crest, often barely visible. Most easily identified in flight by white trailing edge to wings and white sides to tail. Often occurs in large flocks in winter.

Where to see A migrant and winter visitor to the eastern part of the region, favouring open areas of grassland and agriculture.

Crested Lark

Galerida cristata 17cm

Very similar in size and plumage to Eurasian Skylark, but with a distinctive crest which is always obvious. In rather floppy flight note absence of white trailing edge to broad wings and absence of white sides to tail. Usually in pairs and can be very tame. When flushed utters a clear, fluty *du-ee*.

Where to see Resident throughout most of the region, favouring open country, especially cultivation and semi-deserts and often near roadsides and human habitation.

emminck's Lark *Eremophila bilopha* 14cm

ery distinctive black-and-white face attern and black breast-band, features shares with the similar Horned Lark *Eremophila alpestris*, not illustrated) ound in countries bordering the Mediterranean. Note the sandy pperparts of Temminck's Lark and he black cheeks not joined to the reast-band.

Where to see Resident in flat stony and sandy deserts, mainly in the north-west of the region, but also in Mediterranean countries, where Horned Lark breeds on mountain summits.

Greater Short-toed Lark
Calandrella brachydactyla 14cm

A small, rather sandy-coloured lark with streaked upperparts and rather stout bill. The best distinguishing feature from other similar small larks is the black patches at the sides of the neck on otherwise unstreaked underparts. In winter often gathers in large flocks.

Where to see Cultivated fields and semi-deserts throughout the whole region on migration and in winter; also breeds at a few areas near the Mediterranean coast and eastern Arabia.

Bimaculated Lark
Melanocorypha bimaculata 16cm

A large lark that can be confused with the even larger Calandra Lark, but note the subtly different head pattern and, in flight, grey-brown (not blackish) underwing without white trailing edge, and white tip to tail.

Where to see Cultivation, steppe and grasslands, occurring on migration and in winter in the north of the region and eastern Arabia south to Oman. Also breeds in countries bordering the Mediterranean and near the Gulf.

Calandra Lark

Melanocorypha calandra 20cm

One of the largest larks in the region, note its stout bill and black patches on sides of neck. Similar to Bimaculated Lark but distinguished in flight by the white trailing edge to the wings and white sides to the tail.

Where to see Favours cultivated plains, steppe and wastelands. Found only in the north of the region where it breeds in countries bordering the Mediterranean, from which it disperses in winter.

Lesser Short-toed Lark

Alaudala rufescens 13cm

A small, sandy lark that is quite difficult to identify. Closely resembles Greater Short-toed Lark, but smaller with a stouter bill and streaked breast with absence of black neck-patches. Gathers in large flocks in winter.

Where to see Found on migration and in winter throughout the region on steppe, cultivated fields and coastal flats. Breeds semi-colonially at scattered localities in the north and west of the region.

White-eared Bulbul

Pycnonotus leucotis 18cm

Very easily identified by its black head with large white cheek-patches, a feature shown by no other bulbul in the region. Note also the yellow under the tail. It hybridises with the Red-vented Bulbul, producing offspring that are a mix of their plumages. Highly vocal like all bulbuls.

Where to see Native to Iraq, but introduced into Arabia (from Iraq, Iran and Pakistan) and now resident and spreading in northern Syria, along the Gulf coast and in a few areas of Arabia. Found in woodlands, gardens and parks.

Red-vented Bulbul

Pycnonotus cafer 22cm

The black, tufted head and black, scaly breast are the main clues to identification, but note especially the scaly upperparts and red under the tail. In flight shows a whitish rump and white tips to the tail. Gregarious and mixes with other bulbuls.

Where to see Introduced into the region and resident at a few scattered localities along the Gulf south to northern Oman, where it inhabits parks and gardens.

White-spectacled Bulbul *Pycnonotus xanthopygos* 19cm

noisy, gregarious bulbul whose loud, ty *bul-bul-bul* calls are often the first dication of its presence. Rather drab umage with a sooty-black head and oticeable white eye-ring. Note also e yellow under the tail.

Where to see Resident, and the only native bulbul in Arabia, found commonly in countries bordering the Mediterranean, and western and southern Arabia, including Oman and UAE. Frequents any areas with trees, especially those that bear fruit.

Sand Martin
Riparia riparia 12cm

Instantly distinguished from all other swallows and martins, with which it often mixes, by its white underparts with a brown breast-band. Usually seen in flocks on migration and many thousands can gather to roost in a wetland reedbed.

Where to see Widespread in the region on spring and autumn migration when it can be seen especially over wetlands. Breeds at a few colonies in Iraq, excavating nest-holes in sandbanks.

Barn Swallow *Hirundo rustica* 16cm (including long tail)

The dark blue upperparts, deeply forked tail and chestnut throat readily identify this familiar migrant, which often gathers in flocks and mixes with other swallows and martins on migration.

Where to see Widespread over any area on spring and autumn migration, also in winter, except for the north of the region. Breeds in Iraq, Kuwait and countries bordering the Mediterranean, nesting on ledges in buildings.

ale Crag Martin
yonoprogne obsoleta 13cm

n all grey-brown martin that is
ery similar to Eurasian Crag Martin
tyonoprogne rupestris, not illustrated)
hich also occurs in the region, but
ostly on migration. Pale Crag Martin
ffers in being slightly smaller and
aler with clean, off-white underparts,
cluding the chin (streaked in
urasian Crag).

Where to see Resident in gorges,
esert areas and even towns in western
nd southern Arabia and on the Gulf,
cluding Oman.

Common House Martin *Delichon urbicum* 13cm

his 'black-and-white' martin is easily
dentified by its white underparts and
arge white band across its rump. It
ends to fly higher than other swallows
nd martins in its search for insects and
can gather in large flocks on migration.

Where to see A summer visitor to
breeding areas in countries bordering
the Mediterranean and Iraq, and builds
a cup-shaped nest under eaves or on
a cliff face. Otherwise, a widespread
spring and autumn migrant.

Red-rumped Swallow

Cecropis daurica 17cm
(including long tail)

Most closely resembles Barn
Swallow but identified from below
by its black undertail and vent, and
from above by its rufous collar and
pale rufous rump. Usually seen in
pairs on breeding grounds.

Where to see Summer visitor to
breeding areas in countries bordering
the Mediterranean, northern Iraq
and south-west Arabia, building a
flask-shaped nest under a bridge
or on a building. Otherwise, a
fairly widespread spring and
autumn migrant.

Streaked Scrub Warbler *Scotocerca inquieta* 11cm

Small warbler which, with
its long tail, can be confused
with Graceful Prinia, but
distinguished by its dark eye-
stripe bordered by a white
supercilium and finely streaked
breast. Largely ground-dwelling,
rather shy and habitually waves
its long, graduated tail.

Where to see Resident,
generally found in rather
barren areas with low scrub
in countries bordering the
Mediterranean, western and
parts of southern Arabia
and Oman.

Willow Warbler
Phylloscopus trochilus 11cm

A small, active warbler, very similar to Common Chiffchaff, but more olive-green above with a yellow wash to the underparts, a more distinct, yellowish supercilium, longer wings and flesh-coloured legs. In autumn young birds are particularly yellow below.

Where to see A spring and autumn migrant throughout the region where it can be found in any areas with trees and scrub, including desert oases.

Common Chiffchaff *Phylloscopus collybita* 11cm

Very similar to Willow Warbler, but duller, more olive-brown in colour and with shorter wings. Note also dark legs and rather characteristic flicking of its tail as it moves through vegetation searching for food. Its repeated *chiff-chaff* song is commonly heard where breeding and occasionally on migration.

Where to see Can be found throughout the region on migration and in winter in almost any habitat with trees and bushes. Also breeds in woodland in countries bordering the Mediterranean.

Basra Reed Warbler

Acrocephalus griseldis 15cm

One of the region's specialities, being globally Endangered and with virtually the entire world population breeding in the marshes of southern Iraq. Similar in plumage, but larger than Eurasian Reed Warbler and with a noticeably longer bill and more pronounced supercilium.

Where to see A summer visitor to the reedbeds in marshes of southern Iraq, with occasional breeding in nearby countries. More widespread on migration, but nowhere is it commonly seen.

Great Reed Warbler

Acrocephalus arundinaceus 15cm

A very large warbler only confusable with Clamorous Reed Warbler. Much larger than the more familiar Eurasian Reed Warbler with a powerful bill and noticeable supercilium. Very loud, grating, repetitive song indicates its presence on breeding grounds.

Where to see Breeding summer visitor, nesting in scattered colonies in reedbeds in countries bordering the Mediterranean, Iraq and the northern coast of the Gulf. Widespread on spring and autumn migration throughout the region.

Clamorous Reed Warbler

Acrocephalus stentoreus 18cm

A very large warbler, very similar to Great Reed Warbler, but close views show that it has a longer bill, shorter wings and more rounded tail, although these details are subtle. The song is more melodious than that of Great Reed and this, together with its typical mangrove habitat, are the best clues to identification.

Where to see A breeding resident in reedbeds and mangroves mainly on the Mediterranean, Red Sea and Gulf coasts, and also in Oman.

Sedge Warbler

Acrocephalus schoenobaenus 13cm

Small, wetland warbler, identified by streaked buffish upperparts and a noticeable buffish-white supercilium. The other similar warbler in the region is Moustached Warbler (*Acrocephalus melanopogon*, not illustrated) which breeds patchily in Iraq, Mediterranean countries and on the Gulf coast. Moustached differs in having a clear-cut white supercilium, dark crown and rufous-brown upperparts, also often cocks and flicks the tail.

Where to see Found throughout the region on spring and autumn migration in wetland habitats, particularly reedbeds.

Eurasian Reed Warbler *Acrocephalus scirpaceus* 13cm

A small, brown, rather secretive wetland-dwelling warbler that is familiar during times of migration. It resembles Marsh Warbler (*Acrocephalus palustris,* not illustrated), which also occurs on migration, but is warmer brown with a longer bill and shorter wings.

Where to see Found throughout the region on spring and autumn migration in wetland habitats, particularly reedbeds. Also breeds patchily in countries bordering the Mediterranean, on the Gulf coast and at isolated wetlands in Arabia.

Eastern Olivaceous Warbler *Iduna pallida* 12.5cm

Small olive-brown warblers can be difficult to identify. Eastern Olivaceous is one of the commonest and is mostly seen on migration. Note its pale wing-panel, rather long, broad bill and frequent downward movements of its tail.

Where to see Found throughout the region on spring and autumn migration in any areas with trees or scrub. Also breeds in countries bordering the Mediterranean and patchily on the Gulf coast and in central Arabia.

Graceful Prinia *Prinia gracilis* 11cm

very small warbler with a
ong, graduated tail, which is
ften fanned. In suitable habitat
can be common. Fairly tame.
he rather pale face makes
ne reddish-brown eye quite
rominent. Monotonous song is
asy to recognise.

Where to see Resident in
ountries bordering the
Mediterranean and the Gulf,
nd in Oman and western and
outhern Arabia. It is found in
crub and low vegetation, often
lose to human habitation.

Eurasian Blackcap

Sylvia atricapilla 14cm

A medium-sized grey-brown warbler
that is easily identified by its black
cap (male) or brown cap (female and
juvenile). It often calls a persistent *tack*
and, on its breeding grounds, has a
rich, liquid song.

Where to see Widespread on spring
and autumn migration throughout
the region, occurring in almost any
area with trees or bushes. A small
population breeds in woodland on the
Mediterranean coast.

Lesser Whitethroat *Curruca curruca* 13.5cm

Similar to Common Whitethroat, but easily distinguished by dark ear-coverts, dark legs and absence of chestnut in the wing. Migrants from Europe are widespread, but the paler race from Central Asia is found only in winter in southern and central Arabia, favouring acacias in semi-deserts – note its buzzing *che-che-che-che* call.

Where to see Widespread on spring and autumn migration and in winter in any areas with trees and scrub. Breeds in countries bordering the Mediterranean.

Barred Warbler

Curruca nisoria 15.5cm

A rather large, greyish warbler, which can be very skulking. In spring the crescent-shaped bars on the underparts and yellow eye of the adults are important features for identification. In autumn generally greyish, but note white tips to tail, pale edging to wing feathers and barring on undertail-coverts.

Where to see Widespread on spring and autumn migration throughout the region, but generally uncommon, favouring any areas with scrub and thorny bushes.

astern Orphean Warbler *Curruca crassirostris* 15cm

fairly large greyish warbler which resembles a larger version of Sardinian or Arabian Warblers, but male distinguished by black being confined to ear-coverts (not whole head), and yellow eye. Beautiful, musical song delivered from dense cover.

Where to see Fairly widespread on spring and autumn migration, except for southern Arabia, favouring any areas with scrub, bushes and trees. Breeds in countries bordering the Mediterranean, including in orchards and olive groves.

Arabian Warbler
Curruca leucomelaena 14.5cm

This Middle East endemic species is similar to Eastern Orphean Warbler but has a longer black tail with only very thin white outer fringes, and dark eyes. Note especially the characteristic downward and circular movements of the tail when perched.

Where to see Resident in countries bordering south Mediterranean coast, western and southern Arabia, including southern Oman. Occurs in semi-deserts with acacias as well as bush-covered rocky hillsides.

Asian Desert Warbler *Curruca nana* 11.5cm

A small, rather secretive sandy-grey warbler that is usually found in low vegetation or on the ground in arid areas. Often seen following a Desert Wheatear. Note its yellow eyes and rufous rump and tail, which is often flicked and held half-cocked.

Where to see A spring and autumn migrant and also winter visitor to much of the region, but rare in areas bordering the Mediterranean, favourin deserts and semi-deserts with low vegetation.

nis medium-sized warbler
most similar to Lesser
Whitethroat and best
distinguished from it by larger
ze, chestnut fringes to its
ving feathers, orange (not
lack) legs and, in adult, white
ye-ring. Fairly skulking and,
where breeding, note its short,
cratchy song, often uttered in
ong-flight.

Where to see Widespread on
pring and autumn migration
n almost any area with trees,
bushes and especially low
scrub. Breeds in countries
bordering the Mediterranean.

Sardinian Warbler

Curruca melanocephala 13cm

Found only in the Mediterranean
fringe, the male of this small
warbler is easily identified by its
all-black head with contrasting
white throat, grey upperparts and
red eye-ring. Females and young
birds are less distinctive. Very
skulking and often located by its
hard, rattling, call.

Where to see Resident, with
some dispersal, in the countries
bordering the Mediterranean,
where it frequents dry scrubby
areas as well as oak and pine
woodland.

♂

Menetries's Warbler *Curruca mystacea* 12.5cm

Resembles Sardinian Warbler, but male distinguished by black forehead and ear-coverts, which merge into grey crown (distinct from Sardinian's all-black head), and salmon wash to underparts. Female and young birds pale, plain sandy-brown. Constant waving of tail is a good clue to identification.

Where to see Widespread on spring and autumn migration throughout region with some wintering in Arabia; breeds at scattered localities in the north. Occurs in scrub, thickets and often near water.

Abyssinian White-eye
Zosterops abyssinicus 12cm

Small, greyish-green and warbler-like, this white-eye is easily identified by its conspicuous white eye-ring. It is usually seen in pairs or small groups and is very active, constantly on the move and calling a high-pitched *sooee*. Oriental White-eye (*Zosterops palpebrosus*, not illustrated) is found only in Oman, on the island of Mahawt near Barr Al Hikman, and is best identified by its bright yellow forehead and throat.

Where to see Resident in wooded hills in south-west Arabia and southern Oman.

rabian Babbler *Argya squamiceps* 26cm

large, grey-brown, thrush-sized
rd with a long, graduated tail. It is
oisy, mostly ground-dwelling and
sually seen in groups. A bird of the
rabian Peninsula, in Iraq and Syria it
replaced by the similar Iraq Babbler
Argya altirostris, not illustrated) in
etlands and Afghan Babbler (*Argya*
uttoni, not illustrated) in dry areas.

Where to see A resident of the
southern Mediterranean countries,
western and southern Arabia, Oman
and UAE, mostly found in dry scrubby
places, particularly those with
acacia trees.

Western Rock Nuthatch *Sitta neumayer* 14cm

A rock-dwelling nuthatch with pale blue-grey upperparts, whitish underparts and a black eye-stripe. Three nuthatches occur in the region. The tree-dwelling Eurasian Nuthatch (*Sitta europaea,* not illustrated) has rusty underparts and is confined to northern Syria and northern Iraq, whilst Eastern Rock Nuthatch (*Sitta tephronota,* not illustrated) of northern Iraq is similar to Western but is larger, with a heavier bill and extensive black eye-stripe.

Where to see Resident in countries bordering the Mediterranean and northern Iraq, in rocky and boulder-covered areas.

Bank Myna

Acridotheres ginginianus 21cm

Similar to Common Myna, it can be easily identified by its all-grey body with distinctive orange eye-patch and pale red bill. In flight note its rusty wing patches. A noisy bird, often seen in pairs and also gathers in small flocks.

Where to see A non-native species in the region, originating from escapes from captivity and now a common feature of several towns and villages along the Gulf and in Oman.

ommon Myna
ridotheres tristis 23cm

ffers from Bank Myna, with which
often associates, in its vinous-brown
dy, yellow eye-patch and bill. In
ght note white wing patch and
il-corners. Gregarious and noisy
ith loud, liquid, imitative and
petitive song.

here to see A non-native species in
e region, originating from escapes
om captivity. Very common in
wns, villages and parks along the
ulf, along the Red Sea in western
audi Arabia and in Oman.

Common Starling
Sturnus vulgaris 21cm

In winter plumage, when most
likely to be seen in the region,
its greenish-black plumage
is speckled with white spots.
Feeds mainly on the ground
and often gathers in flocks.
Reedbeds are a favourite
roosting place.

Where to see A spring, autumn
and winter visitor to the north
of the region, eastern Arabia
and Oman, occurring in
towns, villages and any open
agricultural country with trees.
Breeds in UAE.

win.

Rosy Starling *Pastor roseus* 21cm

In spring, when migrant flocks pass through the region, the adult is striking in its pink-and-black plumage. In autumn and winter the pink colour becomes greyish-buff but the bird is still very distinct. Juveniles are pale grey-brown with a paler rump, dark wings and yellow bill.

Where to see A spring and autumn migrant to countries in the north of the region, eastern Arabia and Oman, occurring in any open agricultural country with trees.

'istram's Starling *Onychognathus tristramii* 25cm

large, black starling with a riking orange wing patch, ry conspicuous in flight. The ale is glossy black, whilst the male is sooty-coloured with grey head. Often in noisy ocks, note the loud whistling alls which often first attract tention.

Vhere to see Resident in a ariety of habitats, especially ocky hills, semi-deserts nd towns and villages in outhern Levant, and western nd southern Arabia east to outhern Oman.

♀ ♂

♂

♀

Common Blackbird

Turdus merula 25cm

This familiar European bird is found only in the north of the region. Male easily identified by its all-black plumage and yellow bill, whilst females and immatures are brown with brownish bills. The rich, fluty song of the male is wonderful to hear when breeding.

Where to see Winter visitor to north of region south to the Gulf, occurring anywhere with trees and scrub, also cultivation. Breeds in woodland in countries bordering the Mediterranean.

Song Thrush

Turdus philomelos 22cm

Smallest and most widespread of four species of thrushes with spotted breasts in the region. The others (not illustrated) are winter visitors to extreme north: Mistle Thrush *Turdus viscivorus* is the largest, Redwing *Turdus iliacus* has a white supercilium and red flanks, whilst Fieldfare *Turdus pilaris* has a greyish head and lower back.

Where to see A winter visitor to much of the region, but very rare in southern Arabia. It favours areas with trees and bushes, including parks and fields.

Black Scrub Robin

Cercotrichas podobe 18cm

Very easily identified by its black plumage and very long tail, which is white-tipped and often swept up over its back and slightly fanned. Usually skulking and spends much time feeding on the ground, but its melodious song is delivered from a bush top.

Where to see A resident of desert fringes and scrub in western Saudi Arabia and Yemen, and also in isolated areas of central Arabia.

skulking bird that somewhat
sembles a Common Nightingale
shape. However, it has a long
fous tail which is often cocked
nd spread, displaying white tips to
black sub-terminal band.

here to see A spring and autumn
igrant to most of the region,
reeding in countries bordering the
lediterranean and in Syria, Iraq,
ulf coastal areas and Oman. A
ird of semi-deserts and cultivation,
ncluding gardens with scrub.

Spotted Flycatcher *Muscicapa striata* 14cm

This brownish-grey flycatcher is best identified by the dark streakings on its head and breast. Its stance and actions are also characteristic, perching upright on an exposed branch, flicking its wings and making sallies into the air for flying insects.

Where to see Widespread on spring and autumn migration throughout the region, occurring anywhere with trees. Breeds in countries bordering the Mediterranean, in gardens and parks.

Common Nightingale *Luscinia megarhynchos* 16cm

Very secretive and most likely to be heard, rather than seen, in those areas where it breeds. When seen well it is rich brown in plumage with a rusty-red tail and is often on the ground. The similar Thrush Nightingale (*Luscinia luscinia*, not illustrated), a migrant throughout the region, has mottling on the breast.

Where to see Spring and autumn migrant to most of region, breeding in countries bordering the Mediterranean and in north-east Iraq. Occurs in woodland and scrub.

Bluethroat *Luscinia svecica* 14cm

The shape and size of a European Robin, in spring males are readily identified by their blue throat and breast, some with a white or red central spot, varying according to race. Some blue is retained in winter but is lacking in females and immatures; in these, note red sides to the tail in flight. Very skulking.

Where to see Fairly widespread migrant and winter visitor, mostly found in wetlands with reedbeds.

♂

White-throated Robin

Irania gutturalis 16cm

The handsome male is easily identified by its orange-red underparts, with a white throat bordered by black sides to the face. The female is less distinct, but note its orange flanks, white throat and dark brown tail. Skulking and often on the ground.

Where to see Occurs on spring and autumn migration throughout the region in areas with trees and scrub; also breeds in countries bordering the Mediterranean and in northern Iraq, on stony hillsides with shrubby vegetation.

♀

♂

emicollared Flycatcher
cedula semitorquata 13cm

♂

ne most widespread of three very
milar black-and-white flycatchers
at migrate through the region.
ollared Flycatcher (*Ficedula
bicollis*, not illustrated) has a
road white neck-collar, whilst Pied
lycatcher (*Ficedula hypoleuca*, not
lustrated) has no suggestion of a
ollar – both rarely occur in Arabia.
he females are difficult to separate.

Where to see Occurs on spring and
utumn migration in areas with trees
n much of the region except southern
Arabia, but is uncommon.

♀

Red-breasted Flycatcher
Ficedula parva 12cm

The smallest flycatcher to occur in
the region. In spring the orange-red
throat of the male is distinctive,
but females, autumn males and
immatures are best identified by a
black tail with conspicuous white
sides. A perky, active bird that often
perches with tail cocked.

Where to see A migrant to the
region and can occur anywhere with
trees in the countries bordering the
Mediterranean; also occurs in the
Gulf countries and Oman, where it
also winters.

Robins, thrushes and flycatchers

Black Redstart *Phoenicurus ochruros* 15cm

Likely to be confused only with Common Redstart, but note much darker grey plumage with more extensive black on underparts of the male. In some races the black is clearly demarcated from the red lower breast and belly. Actively shivers rusty-red tail, especially when landing.

Where to see Widespread on migration and in winter in almost any area, including open woodland, rocky slopes, towns and villages. Breeds in mountains close to the Mediterranean and in northern Iraq.

Common Redstart

Phoenicurus phoenicurus 15cm

Unlike Black Redstart, rarely found in winter. A male in spring is identified by its grey back, black throat, red underparts and white forehead. One race shows an obvious white wing-panel. Females and immatures similar to female Black Redstart but noticeably paler and much browner. Actively shivers rusty-red tail.

Where to see Widespread on spring and autumn migration, occurring in any area with trees. Breeds in countries bordering the Mediterranean and in northern Iraq.

Common Rock Thrush

Monticola saxatilis 19cm

The male of this largely ground-dwelling thrush-like bird is easily identified by the white patch on its blue-grey upperparts, blue-grey bib and red underparts and tail. The female is similar to female Blue Rock Thrush but note white scaling on upperparts and red tail.

Where to see Widespread on migration where found in most habitats; rare in winter (unlike Blue Rock Thrush). Breeds in rocky mountains close to the Mediterranean Sea.

Blue Rock Thrush

Monticola solitarius 21cm

Similar in shape to Common Rock
Thrush but slightly larger. Male
easily identified by its dull, dark blue
plumage. Female and young birds are
very similar to Common Rock Thrush
but lack white scaling on upperparts
and have a brown (not red) tail.

Where to see Widespread on
migration and in winter throughout
the region. Breeds in rocky mountains
and on cliffs and ruins in countries
close to the Mediterranean and in
northern Iraq.

Whinchat

Saxicola rubetra 13cm

An upright chat that often perches
on the top of low vegetation. Male
in spring easily told by its clear-cut
white supercilium, dividing blackish
cheeks and crown. Female and
immature less distinctive, but always
show a white or buff supercilium and
white sides to base of tail, noticeable
in flight.

Where to see Occurs on spring
and autumn migration throughout
the region in any open country
with bushes.

European Stonechat

Saxicola rubicola 12cm

Small chat that often perches in an exposed position on top of low vegetation. Male very similar to male Siberian Stonechat and best told by its brownish, streaked rump (white or salmon in Siberian). Females and immatures resemble Whinchat but are darker, lack white supercilium and have an all-black tail.

Where to see Winter visitor to countries bordering the Mediterranean, in Iraq and along the coast of the Gulf, favouring any open country with bushes.

Siberian Stonechat

Saxicola maurus 12cm

Very similar to European Stonechat and, like that species, often perches in an exposed position on top of low vegetation. Male best told by its broad white or salmon-coloured rump. Females and immatures are noticeably paler than European Stonechat with a pale rump and narrow off-white supercilium.

Where to see Winter visitor to much of the region except western Arabia, occurring at a range of altitudes in any open country with shrubs.

Northern Wheatear

Oenanthe oenanthe 15cm

An active ground-dwelling bird in a family where many species are similar in female and immature plumages. Males in spring are told by their grey back and black eye-mask; the buff-brown females and immatures are less distinct and best separated from other wheatears by their black tail-band of even width.

Where to see Widespread spring and autumn migrant throughout region in any open area. Breeds in mountainous regions of countries bordering the Mediterranean.

♂

♀

Isabelline Wheatear *Oenanthe isabellina* 16cm

Large, upright-standing wheatear similar to large female Northern Wheatear, and some individuals can be difficult to distinguish. Note more sandy plumage, especially wings, black line from bill to eye and, in flight, broader black tail-band with small T-shaped projection into white rump.

Where to see Widespread spring and autumn migrant and winter visitor throughout the region in open country. Breeds in countries bordering Mediterranean, across Syria to northern Iraq, in steppe and grassy areas.

Desert Wheatear
Oenanthe deserti 15cm

A sandy wheatear best told at all ages by all-black tail, contrasting with white or buffish-white rump. The handsome male has striking sandy plumage with black face and wings; female and immature birds lack black face. Curiously, males outnumber females up to ten to one on wintering grounds.

Where to see Widespread on migration and in winter throughout region, occurring in any open habitat, particularly semi-deserts. Breeds in countries bordering the Mediterranean.

Pied Wheatear *Oenanthe pleschanka* 15cm

Males of several wheatears in the region are black and white and care is needed for identification. Male Pied most resembles Eastern Black-eared Wheatear but has extensive black on the back and sides of face which connects to the black wings. Females and immatures are very similar to Eastern Black-eared and much experience is needed to identify them.

Where to see Spring and autumn migrant throughout the region, occurring in any open area with scattered vegetation.

Eastern Black-eared Wheatear *Oenanthe melanoleuca* 14cm

The male of this handsome wheatear has two forms: one with a black throat and one with a black eye-mask. Note in black-throated form the black does not join with the black wings, important for distinguishing from Pied Wheatear. Females and immatures difficult to separate from Pied.

Where to see Spring and autumn migrant throughout region, occurring in any open area. Breeds on stony, sparsely vegetated slopes in countries bordering Mediterranean and in northern Iraq.

◀ ackstart

enanthe melanura 15cm

familiar bird of western Arabia here easily told by its grey umage and black tail. Often first oticed – as it sits prominently n a low bush – by its active novements, lowering and preading its tail, whilst lowering s wings.

Where to see Resident in countries bordering the southern Mediterranean coast south along he Red Sea coast to Yemen and outhern Oman, occurring in dry, parsely vegetated areas.

◀ White-crowned Wheatear *Oenanthe leucopyga* 17cm

Large wheatear with all-black plumage except for white rump, white lower underparts and – usually – white crown. In flight note extensive white sides to tail. Male Hooded Wheatear (*Oenanthe monacha*, not illustrated) has a similar pattern but its underparts, below a black bib, are white; however, this species is rare and has a patchy distribution in Arabia.

Where to see Resident in the rocky deserts and mountains of western and central Arabia, usually without vegetation and often near human settlements.

Hume's Wheatear *Oenanthe albonigra* 16cm

Similar to Variable Wheatear (*Oenanthe picata*, not illustrated), which has various forms and occurs, only in winter, in the same area. Hume's is larger, more bull-headed and, most noticeably in flight, the white on its back extends further up between the wings – cut-off and lower in Variable Wheatear. Note also glossy black plumage compared to sooty-black of Variable.

Where to see Resident in barren, stony habitats from plains to mountain slopes in UAE and northern Oman.

Finsch's Wheatear *Oenanthe finschii* 15cm

Resembles both Eastern Black-eared and Pied Wheatears, but separated by a combination of black on face and throat joining solidly to black wings, and the buffish-white on the back extending between the wings to the white rump – especially noticeable in flight.

Where to see Winter visitor to the north of the region and the countries bordering the Gulf, with breeding populations in northern Iraq. Favours dry rocky areas and barren, rocky slopes.

Wheatears

Mourning Wheatear *Oenanthe lugens* 14cm

Black-and-white wheatear that requires care to identify. Most similar to Pied Wheatear but told by the orange wash under the tail and, in flight, by the white panels in the wing. The nearly all-black Basalt Wheatear (*Oenanthe warriae*, not illustrated) occurs in the black-stone deserts of Jordan; until recently it was considered a race of Mourning Wheatear.

Where to see Breeds on barren slopes and stony semi-deserts in countries bordering the Mediterranean and in northern Saudi Arabia; otherwise uncommon in winter in the Gulf countries.

Arabian Wheatear
Oenanthe lugentoides 14cm

Very similar to Mourning Wheatear but the two species do not overlap in range. Like Mourning Wheatear, it has an orange wash on the undertail, but the white panels in the wing (when seen in flight) are smaller and it usually has a grey-streaked crown.

Where to see Resident in sparsely vegetated rocky hills and mountains in southern Arabia, extending to southern Oman.

oth sexes have a rather dull
uffish-grey plumage but
an be instantly told by their
range-red rump and sides to
e tail, which are very obvious
n flight. Kurdistan Wheatear
Oenanthe xanthoprymna, not
llustrated), which is found in
aq, also has a red rump but its
ail-sides are white.

Where to see A winter visitor
o the Gulf countries and
Oman, occurring in a variety
of open habitats, often without
vegetation, including cultivation,
emi-deserts and ruins.

Nile Valley Sunbird

Hedydipna metallica 10cm
(plus 5cm tail in male)

Adult male in full breeding
plumage is unmistakable with
its glossy green-and-purple
upperparts and breast, yellow
underparts and very long
tail feathers. The plumages
of females, immatures and
males outside the breeding
season resemble those of other
sunbirds but note the small,
short bill.

Where to see Resident in
south-west and southern Arabia,
extending to southern Oman,
in a variety of habitats with
flowering trees and shrubs.

Sunbirds

Palestine Sunbird *Cinnyris osea* 11cm

A small sunbird and the only one occurring in Mediterranean countries. Iridescent blue-black male distinguished from Shining Sunbird, which occurs in the same areas in Arabia, by smaller size, shorter tail, bluer plumage and absence of red breast-band. It also frequently flicks its tail – helpful in separating from female and immature Shining Sunbird.

Where to see Found from Lebanon through western and southern Arabia to south Oman, in well-vegetated areas with flowering trees and shrubs.

Shining Sunbird *Cinnyris habessinicus* 13cm

Larger than the similarly coloured Palestine Sunbird, and adult male identified instantly by its red breast-band. Females and immatures distinguished from Palestine Sunbird by larger size, longer bill and grey, slightly vermiculated underparts. Note that Shining Sunbird has pronounced slow flicking of its broader tail compared to the nervous flicking of Palestine Sunbird.

Where to see Western and southern Arabia to south Oman, in well-vegetated areas with flowering trees and shrubs.

♀

♂

Purple Sunbird

Cinnyris asiaticus 10cm

This small sunbird resembles
Palestine Sunbird but is immediately
identifiable in the UAE and northern
Oman as it is the only sunbird
occurring there. The metallic blue-
blacked male in breeding plumage
can occasionally show a reddish
breast-band. Female and immature
are washed yellow on underparts as
is the male out of breeding plumage,
when it also has a dark centre to the
throat and breast.

Where to see Found only in Oman
and UAE, inhabiting cultivation and
deserts with flowering trees
and scrub.

♀

♂

House Sparrow

Passer domesticus 15cm

The sparrow that is so well known in Europe is also commonly seen in the Middle East. Spanish Sparrow also occurs in the north of the region, but male House Sparrow is instantly identifiable by the lack of black streaking on the underparts. The females and immatures are more difficult to separate. Often found in small flocks.

Where to see Widespread resident in most of the region, in towns, villages and farmland.

♂

♀/imm.

Spanish Sparrow

Passer hispaniolensis 15cm

The same size as House Sparrow, but instantly identified by heavily black-streaked underparts and back as well as brown (not grey) crown. Females and immatures very similar to House Sparrow and best told by their association with males. Nests colonially, often in nest of another bird, such as a White Stork.

Where to see Breeds in countries bordering the Mediterranean, in northern Syria, Iraq and on the Gulf coast, dispersing south in winter to reach Oman.

♂

♀

Dead Sea Sparrow *Passer moabiticus* 15cm

The smallest sparrow in the region. The male is best identified by its head pattern, especially the grey cheeks, long white-and-rusty supercilium and yellow moustache. The female is like a miniature female House Sparrow. Nests colonially and is also gregarious outside the breeding season.

Where to see Resident in countries bordering the southern Mediterranean and patchily in Syria and Iraq, favouring thick scrub, tamarisk and poplars, always near water.

Rock Sparrow
Petronia petronia 15cm

Similar in size to House Sparrow, but readily told by the boldly striped markings on its underparts and broad dark and buff makings on its head. This is the most streaked sparrow in the region. Occurs and nests colonially.

Where to see Resident in rocky mountains and other stony areas, farmland and habitation, especially ruins, in countries bordering the Mediterranean and in northern Iraq.

ellow-throated Sparrow *Gymnoris xanthocollis* 14cm

reyish-olive, rather indistinct
parrow, the male told by
hestnut shoulders, white
ing-bar and, if seen well,
ellow throat-patch. Female
nd immature lack the chestnut
houlder and yellow throat.
ale Rockfinch (*Carpospiza
rachydactyla*, not illustrated) is
arger and similar to female, but
ote white tip to tail.

Where to see Breeding summer
isitor to Iraq, Kuwait, UAE and
northern Oman, mainly in dry
woodland and olive groves. In
winter more widespread in the
Gulf region and Oman.

Rüppell's Weaver
Ploceus galbula 15cm

Probably the first indication of this
colonial weaver's presence will be the
sight of its basket-like nests hanging
in groups from the branches of a tree,
usually an acacia. Male instantly
identified by its yellow plumage and
dark chestnut face. Sparrow-like
females, and winter males, are
olive-brown above and washed
yellow below.

Where to see Resident in south-west
Arabia and southern Oman, favouring
acacia trees in cultivated and dry
areas; introduced to Kuwait.

African Silverbill
Euodice cantans 11cm

Small, slim, finch-like bird, very similar to Indian Silverbill but the two species occur in different areas. Both have a large, conical, 'silver' bill, but African is identified by black rump and uppertail-coverts (whitish in Indian Silverbill). Note also fine vermiculations on upperparts. Fairly tame and often in small, noisy groups.

Where to see Resident in south-west Arabia and southern Oman, occurring in grassland, scrub and cultivation.

Western Yellow Wagtail *Motacilla flava* 17cm

A complex species with many races, each with different head patterns, varying from grey to blue to yellow, the blue-headed races often having a supercilium above the eye. One of the commonest races, Sykes's Wagtail *Motacilla flava beema*, is illustrated. On spring migration, large flocks with a mix of races often occur.

Where to see Widespread throughout the region on spring and autumn migration in wetland margins and both wet and dry fields.

...dian Silverbill *Euodice malabarica* 11cm

...nall, slim and similar to African ...lverbill, but the two species occur ... different areas of the region. Both ...ave a large, conical, 'silver' bill, but ...dian is identified by whitish rump ...nd uppertail-coverts (black in African ...ilverbill); also lacks vermiculations ...n upperparts. Fairly tame and often ...n groups.

Where to see Resident from northern Oman and along the Gulf; also patchily in north-west Arabia, occurring in grassland, scrub and cultivation. Introduced to Kuwait.

Citrine Wagtail

Motacilla citreola 18cm

Similar in shape to Western Yellow Wagtail, but in spring the male is unmistakable with its bright yellow head, black hind-neck collar, grey back and white wing-bars. Less distinctive in winter, but note grey upperparts and broad white wing-bars, also seen in females and immatures.

Where to see Found throughout most of the region on migration and in winter at margins of freshwater wetlands, but generally uncommon, especially in north of region.

♂ win.

♂ sum.

Grey Wagtail
Motacilla cinerea 18cm

The wagtail with the longest tail, which is constantly wagged up and down. A combination of grey upperparts and yellow under the tail are key to identification in all plumages. Male in breeding plumage also has a black throat. Occurs singly or in pairs, never flocks.

Where to see Widespread in the region on migration and in winter, always near water. Breeds in countries bordering the Mediterranean and in northern Iraq on streams in wooded hills.

White Wagtail
Motacilla alba 18cm

This grey, black and white bird is the commonest wagtail in the region in winter. Male in summer plumage has white face with black crown, nape and throat; in winter the latter reduces to just a black necklace, a feature shown by females and immatures.

Where to see Widespread in any open areas, including cultivation and near habitation. Breeds in countries bordering the Mediterranean.

Tawny Pipit *Anthus campestris* 17cm

Pipits are ground-dwelling birds and can be quite tricky to identify as they are all very similar. Tawny Pipit is one of the larger species and best identified by its rather plain sandy colour with a pale, almost unstreaked breast. Runs fast, often stopping suddenly and standing quite upright.

Where to see Occurs on migration and in winter throughout the region where found on any open ground. Breeds in countries bordering the Mediterranean.

Long-billed Pipit *Anthus similis* 17cm

Similar to Tawny Pipit and the two are often confused. Habitat is a clue as Long-billed Pipit is found in mountains whereas Tawny occurs on plains. Plumage-wise, Long-billed Pipit differs from Tawny in having a warm buff wash on the flanks and vent, slight streaking on the breast and buff (not white) outer tail feathers.

Where to see Resident in rocky hills in countries bordering the Mediterranean, in south-west Arabia, and southern and northern Oman.

ee Pipit

nthus trivialis 15cm

small, neatly streaked pipit very
milar to Meadow Pipit (*Anthus
atensis,* illustrated right) and best
parated by voice: in flight Tree Pipit
alls a distinctive *speeze*, whereas
Meadow Pipit utters a thin, *tsip, tsip*.
ther subtle differences include the
inker legs and less streaking on flanks
f Tree Pipit.

Where to see A spring and autumn
nigrant throughout the region where it
an be found in any open countryside.

Meadow Pipit

Pipits

Red-throated Pipit *Anthus cervinus* 15cm

Most similar to Tree and Meadow Pipits, except in summer plumage when immediately distinguished by its reddish face, throat and upper breast, making it quite distinctive. Outside of breeding plumage note the streaked rump when seen in flight, the broad streaks on the flanks and the high-pitched, drawn-out flight-call *pseeeee*

Where to see Throughout the region on migration and in winter in wetland margins and grasslands.

Water Pipit *Anthus spinoletta* 16cm

A medium-sized pipit which differs in all plumages from other pipits found in the region by its dark legs. Birds in summer plumage, which are seen on spring migration, have virtually unstreaked breast washed in rosy-pink. In winter the underparts are white, barely streaked black.

Where to see A migrant and winter visitor to much of the region, except southern Arabia; found in grasslands and the margins of wetlands.

Common Chaffinch

Fringilla coelebs 15cm

Male told by blue-grey crown and nape, contrasting with pinkish face and breast, and noticeable double white wing-bars. The brownish-olive female also shows conspicuous wing-bars. Brambling (*Fringilla montifringilla,* not illustrated), occurring in same area in winter, separated by white rump and absence of white wing-bars.

Where to see Winter visitor to north of the region, barely reaching the Gulf, occurring in fields and scrub. Breeds in woodlands in countries bordering the Mediterranean and in northern Iraq.

Trumpeter Finch

Bucanetes githagineus 14cm

A small finch of arid areas in the Middle East. The stout orange-red bill of the male is distinctive, along with its grey head and the pinkish wash on underparts, wings and rump. Females and immatures are plain sandy-grey. Note orange legs in all plumages.

Where to see Patchily distributed in countries bordering the Mediterranean, in northern and central Arabia, Yemen, Oman and isolated areas on the Gulf; on dry rocky hillsides, stony deserts and semi-deserts.

ommon Rosefinch

arpodacus erythrinus 14cm

small finch with a stout bill. Male
breeding plumage with red head
nd breast is unmistakable, but such
rds are rare in the region whilst
n migration. The more common
lumage of males, females and
nmatures is olive-brown with
reaked underparts and two
ing-bars.

Where to see A migrant and rare
winter visitor to areas along the Gulf
oast south to Oman, occurring
nywhere with trees and scrub.

Sinai Rosefinch

Carpodacus synoicus 14cm

This Middle East speciality is endemic
to the region. Male easily told by soft
salmon-pink head and underparts, the
colour intensified on the face, and
with a lovely silver cast. Females and
immatures are rather nondescript pale
buff with light streaking below and
a ginger wash on the face. Often in
flocks outside breeding season.

Where to see Countries bordering the
southern Mediterranean and the very
north of Saudi Arabia, where resident
in arid rocky hills.

European Greenfinch
Chloris chloris 15cm

A rather stocky finch. The male is olive-green with yellow wing patches and yellow outer tail feathers, features also seen in the much duller female and immature plumages. Note bat-like display flight and loud, canary-like trilling song when breeding.

Where to see Breeds in countries bordering the Mediterranean and in northern Iraq in pine woods, plantations and parks. In winter also found in farmland and occurs further south, but rarely to north Arabia.

Desert Finch
Rhodospiza obsoleta 15cm

This greyish-buff finch is readily told by its pink-and-white wing pattern and stout, black bill. The bill of the male is accentuated by black around its base, which extends to the eye. Frequently feeds on the ground and often in small, loose flocks.

Where to see Resident, with a patchy distribution including countries bordering the Mediterranean, Kuwait and isolated areas in Saudi Arabia; occurs in open country with trees, bushes and orchards.

Arabian Golden-winged Grosbeak *Rhynchostruthus percivali* 15cm

This Middle East speciality is endemic to the region and is classified as Near Threatened. Easily identified by the combination of its large black bill, white cheek-patches and yellow in the wings and tail. Often occurs in small groups, frequently uttering soft, fluty, rippling calls.

Where to see Only found in the hills of south-west Arabia, especially Yemen, and in southern Oman, where it favours acacia trees and euphorbias.

Common Linnet

Linaria cannabina 13cm

Male in breeding plumage identified by grey head, red forehead and breast, which features distinguish it from all other finches in the region. Duller female and immature plumages have white fringes in wing and outer tail feathers. Yemen Linnet (*Linaria yemenensis*, not illustrated) is similar, but note broad white wing-bar in flight.

Where to see Breeds in countries bordering the Mediterranean in open country with trees and bushes. In winter moves south to northern Arabia and the northern Gulf.

European Goldfinch *Carduelis carduelis* 14cm

Both males and females are easily identified by their striking red, white and black head pattern and broad yellow wing-bar. Often occurs in flocks outside the breeding season.

Where to see Breeds in countries bordering the Mediterranean in a wide variety of habitats with trees and shrubs, including woodlands and cultivation. Winters further south to northern Arabia, often in open country and wasteland.

European Serin
Serinus serinus 12cm

A small finch with prominent dark streaking on upperparts and on white underparts. Both sexes have obvious yellow rump, seen in fast, bouncing flight. Male has bright yellow face but female lacks yellow apart from rump. Usually in small flocks outside breeding season.

Where to see Breeds in countries bordering the Mediterranean in woodlands, orchards and bushy areas, spreading eastwards in winter when found in any open country.

Syrian Serin *Serinus syriacus* 12cm

Middle East endemic that is classified as Vulnerable. Identified by yellow face, virtually unstreaked whitish underparts and pale grey upperparts with broad yellow bands on wings in flight. Breeds socially. Found in small flocks in winter and feeds mostly on the ground.

Where to see Breeds in mountains and hills with trees, especially junipers, in countries bordering the Mediterranean southwards from southern Syria. Some dispersal to lower, open areas in winter.

Corn Bunting

Emberiza calandra 18cm

A large, rather plump bunting with a large bill and noticeably streaked underparts. Unlike other streaked buntings it does not have white outer tail feathers. In flight it often dangles its legs. Frequently found in small flocks outside the breeding season.

Where to see Breeds in countries bordering the Mediterranean and across northern Syria and Iraq in open areas with scattered bushes. In winter occurs south to northern Arabia, the Gulf and Oman.

Rock Bunting

Emberiza cia 16cm

Black-and-grey striped head and orange belly of the male (streaked orange in female) makes this bunting easy to identify, except from Striolated Bunting, from which told by streaked back, grey bill and white sides to tail. Largely ground-dwelling.

Where to see Breeds in countries bordering the Mediterranean, in hills and mountains with rocky and bushy slopes. In winter moves to lower areas to the east, but barely into northern Arabia.

Cinereous Bunting

Emberiza cineracea 17cm

This globally Near Threatened medium-sized bunting is subtly plumaged in soft greys and yellow-greens, with female and immature birds duller and more noticeably streaked. Note also the blue bill, obvious pale eye-ring and two wing-bars. Mainly ground-dwelling.

Where to see A breeding summer visitor to sparsely wooded rocky hills in northern Iraq. In winter quite widespread, except for most of Arabia, favouring semi-deserts and vegetated wadis.

Ortolan Bunting

Emberiza hortulana 17cm

Olive-grey head and breast-band of male with a yellow throat and moustachial stripe distinguishes it from most other buntings. Care must be taken, however, with separation from the similar Cretzschmar's Bunting (*Emberiza caesia,* not illustrated, which occurs in rocky hills in Mediterranean areas), which has a blue-grey head and breast-band with rusty-orange (not yellow) throat.

Where to see Breeds in countries bordering the southern Mediterranean in open country with scattered bushes. Widespread in region on migration.

Striolated Bunting *Emberiza striolata* 14cm

Very similar to Rock Bunting but does not occur in the same areas of the Middle East. Note orange bill, rufous shoulder patch and finer streaks on upperparts; it also lacks wing-bars and outer tail feathers are rufous, not white. Fairly shy.

Where to see Jordan, western and southern Arabia, Oman and UAE, where resident in barren, rocky hills and oases with little or no vegetation.

Cinnamon-breasted Bunting *Emberiza tahapisi* 17cm

A striking, fairly large bunting with a bold black-and-white head pattern, all-black throat and rufous-orange underparts – features that are much bolder and brighter than on the smaller Striolated Bunting, whose range it overlaps with in parts of southern Arabia. Ground-dwelling and often found in small flocks.

Where to see Rocky hills and slopes with bushes and scattered vegetation in south-west Arabia and the extreme south of Oman, where abundant.

Black-headed Bunting
Emberiza melanocephala 17cm

Male in breeding plumage unmistakable with black head, chestnut upperparts and yellow underparts, features that are muted in winter. Female is more difficult to identify, but important clues are the warm brown upperparts, two white wing-bars, stout grey bill and yellowish wash below.

Where to see Breeds in countries bordering the Mediterranean and in northern Iraq, mostly in open country with bushes and scattered vegetation. Occurs on migration throughout eastern Arabia, the Gulf and Oman.

Common Reed Bunting *Emberiza schoeniclus* 16cm

win.

The male's smart breeding plumage, with its black head and breast and white moustache, is unlikely to be seen in the region. In winter resembles female and immature and best told by streaked brown upperparts, dark cheeks with pale supercilium and whitish moustache.

Where to see Winter visitor to the north of region and likely to be found then only in countries bordering the Mediterranean, where it frequents wetland habitats, especially reedbeds.

FURTHER READING AND RESOURCES

The Ornithological Society of the Middle East, the Caucasus and Central Asia (OSME)
The Ornithological Society of the Middle East was formed in April 1978 as a successor to the Ornithological Society of Turkey (established in 1968) and expanded in 2001 to cover the Caucasus and Central Asia. The Society promotes ornithology and conservation throughout the region and publishes the highly respected journal, *Sandgrouse*. The work of the Society is entirely supported by its membership in the region and around the world. As well as providing a forum for exchanging information about the birds of the region, OSME also provides small grants to support conservation and youth development projects. It is a UK-based charity that is run entirely by volunteers. You can learn more about the work of OSME by visiting the website: osme.org.

BirdLife International – Middle East Secretariat
The BirdLife Middle East programme started in 1993 with the launch of *Important Bird Areas in the Middle* East, since when a Middle East office was opened in Amman, Jordan, in 1998. The programme now supports and coordinates conservation activities by the BirdLife Partners and Affiliates across the region. Currently the Middle East partnership is active in Iraq, Jordan, Kuwait, Lebanon, Palestine, Saudi Arabia and Syria, and there are plans for expansion in Iran, Oman, UAE and Afghanistan. The main aims of the Secretariat and the Regional Partnership work are to develop and strengthen regional, cross-regional and global communication, enhance the knowledge base on bird populations, threats and sites, determine priorities and facilitate actions for bird conservation. High on its agenda is the monitoring and protection of IBAs, conserving migrating soaring birds on internationally important flyways and tackling illegal bird killing. For further information see: birdlife.org/middle-east and facebook.com/BirdLifeME.

General
BirdLife Data Zone datazone.birdlife.org Country species lists with global status, including full details of all threatened species and helpful country profiles.
Birds of the Middle East by Richard Porter & Simon Aspinall (Helm, English).
Birds of the Middle East (in Arabic) by Richard Porter & Simon Aspinall (translated by AbdulRahman Al-Sirhan) (Helm, also available as a free-to-download app).
Collins Bird Guide by Lars Svensson, Killian Mullarney & Dan Zettersrom (Collins).
eBird ebird.org/region Many birding sites in the Middle East are covered here, with species lists and locations.
International Ornithological Congress list of bird names worldbirdnames.org.

Bahrain
Birds of Bahrain by Tom Nightingale & Mike Hill (Immel Publishing).
Bahrain Natural History Society bahrainwildlife.com.

Iraq
Key Biodiversity Areas of Iraq: Priority Sites for Conservation & Protection by Nature Iraq (Tablet House Publishing).
Nature Iraq natureiraq.org (BirdLife Partner). For details of Qara Dagh eco-lodge, contact Hana Raza of Nature Iraq: hanaaahmad.raza@gmail.com

Israel

Society for the Protection of Nature in Israel birds.org.il (BirdLife Partner).

The Birds Portal of the Israel Ornithological Centre birds.org.il/en (BirdLife Israel) has much useful information including the latest checklist, places to visit, centres and tours.

Jordan

The Birds of the Hashemite Kingdom of Jordan by Ian Andrews (privately published).

Important Bird Areas in Jordan. Free to download: jo.chm-cbd.net/biodiversity

The State of Jordan's Birds (RSCN) datazone. birdlife.org/userfiles/file/sowb/countries/SOJB_final_pdf.pdf

The Royal Society for Conservation of Nature in Jordan rscn.org.jo (BirdLife Partner).

Kuwait

Birds of Kuwait Birdsofkuwait.com. Here you can find the official Kuwait Bird List as endorsed by the Kuwait Ornithological Rarities Committee, a Kuwait Bird Gallery, details of birding sites and regular birding tours.

Kuwait Environment Protection Society keps.org.kw (BirdLife Partner).

Lebanon

Birds of Lebanon: a photographic guide to 404 species, by Ghassan Ramadan-Jaradi & Fouad Itani.

State of Lebanon's Birds and IBAs (Society for the Protection of Nature in Lebanon).

Association for Bird Conservation in Lebanon: birdsoflebanon.com.

Society for the Protection of Nature in Lebanon spnl.org (BirdLife Partner).

Oman

Birds of Oman by Jens Eriksen & Richard Porter (Helm).

Common Birds of Oman (3rd edition) by Hanne & Jens Eriksen (Al Roya Publishing).

Birdwatching guide to Oman by Dave E Sargeant and Hanne & Jens Eriksen (Al Roya Publishing).

Birds Oman birdsoman.com. Latest Oman Bird List and other birdwatching information.

Environment Society of Oman eso.org.om.

Al Ansab Wetland reservation information haya.om/en/Pages/Wetland.aspx.

Palestine

Palestine Wildlife Society wildlife-pal.org (BirdLife Partner).

Qatar

Birds of Qatar qatarbirds.org. The website of the Qatar Bird Records Committee, which maintains the national list.

Friends of the Environment Center fec.org.qa.

Saudi Arabia

Birds of Saudi Arabia (2 volumes) by Chris Boland & Abdullah AlSuhaibany (Motivate, Dubai).

Field Guide to the Biodiversity of Dhahran by Chris Boland, Jem Babbington, Phil Roberts & I Linning (Motivate, Dubai).

Birds of Saudi Arabia birdsofsaudiarabia.com. Website by Jem Babbington on birds and wildlife throughout the Kingdom.

Syria

Syrian Society for the Conservation of Wildlife sscw-syr.org (BirdLife Partner).

United Arab Emirates

Birds of the United Arab Emirates by Simon Aspinall & Richard Porter (Helm).

Emirates Natural History Group enhg.org/. A long-standing group with chapters in several parts of the country.

UAE Birding. uaebirding.com. Locally run website packed with information, including site details for Abu Dhabi Island.

PHOTO CREDITS

Bloomsbury Publishing would like to thank the following for providing photographs and permission to use copyright material.

Key: T = top; C = centre; B = bottom; L = left; R = right; BL = bottom left; BR = bottom right; TL = top left; TR = top right; CL = centre left; CR = centre right.

AR = AbdulRahman Al-Sirhan; HJE = Hanne and Jens Eriksen.

Front cover TL HJE, TC AR; TR AR; B HJE; **Back cover** T AR, C AR, B HJE, B (Large) HJE; **1** JE; **3** AR; **8** Julian Baker; **9** L Abdulla Al Kaabi, R Abdulla Al Kaabi; **10** T Hana Ahmad Raza, B Richard Porter; **11** TL Jonathan Meyrav, TR Jonathan Meyrav; **12** T Nashat Hamidan, B Nashat Hamidan; **13** T AR, C AR, B Samer Azar; **14** HJE; **15** T HJE, B Imad Atrash; **16** L Imad Atrash, R Gavin Farnell; **17** L Mansur Al Fahad, R Jem Babbington; **18** T Issam Al-Hajjar, CL Issam Al-Hajjar, BL Issam Al-Hajjar; **19** L Oscar Campbell, R Oscar Campbell; **20** L Oscar Campbell, R Richard Porter; **21** T HJE, B HJE; **22** TR AR, CR AR, CL HJE, BL HJE; **23** TL AR, TR AR, B HJE; **24** T AR, B HJE; **25** T AR, B HJE; **26** T HJE, C HJE, B HJE; **27** T AR, B HJE; **28** TR HJE, CR AR, BL HJE; **29** TR HJE, CR HJE, BL AR, BR AR; **30** TR HJE, CR AR, BL AR; **31** T HJE, C HJE, B HJE; **32** TL HJE, TR HJE, B HJE; **33** T HJE, BL AR, BR AR; **34** TR HJE, CR AR, CL HJE, BL HJE; **35** T AR, B AR; **36** T AR, B HJE; **37** TR AR, CR AR, BL HJE; **38** T HJE, B AR; **39** TR HJE, CR HJE, B HJE; **40** T HJE, B HJE; **41** T AR, B AR; **42** T AR, BL AR, BR HJE; **43** T AR, BL HJE, BR HJE; **44** T AR, B HJE; **45** T HJE, B HJE; **46** T HJE, B HJE; **47** TL HJE, TR AR, BL AR, BR AR; **48** T AR, B HJE; **49** TR AR, CR HJE, CL AR, BL AR; **50** TL HJE, TR HJE, BL HJE, BR AR; **51** TR HJE, CR HJE, CL AR, BL AR; **52** TL HJE, TR HJE, B AR; **53** T HJE, BL AR, BR AR; **54** T AR, B HJE; **55** T HJE, CL HJE, BL HJE; **56** TR HJE, CR HJE, B HJE; **57** TR HJE, CR HJE, CL HJE, BL HJE; **58** T HJE, B HJE; **59** TL HJE, TR HJE, BL HJE, BR AR; **60** TR HJE, CR AR, CL HJE, BL HJE; **61** CR AR, CL HJE; **62** T AR, BL AR, BL AR; **63** T AR, B HJE; **64** T HJE, C HJE, B HJE; **65** T AR, C HJE, B HJE; **66** TR HJE, CR HJE, CL AR, BL HJE; **67** T HJE, B HJE; **68** T AR, B AR; **69** T HJE, B HJE; **70** TR AR, CR HJE, CL AR, BL HJE; **71** TR HJE, CR HJE, CL AR, BL AR; **72** T HJE, CR AR, B AR; **73** AR; **74** TR HJE, CR HJE, CL HJE, BL HJE; **75** T AR, C HJE, B AR; **76** T AR, B AR; **77** T HJE, C AR, B AR; **78** T HJE, B HJE; **79** HJE; **80** TL HJE, TR AR, BL AR, BR AR; **81** T HJE, C HJE, B HJE; **82** T HJE, B HJE; **83** TL HJE, TR AR, B HJE; **84** T HJE, B AR; **85** TR AR, CR HJE, CL HJE, BL HJE; **86** TR HJE, CR HJE, BL AR, BR AR; **87** TL HJE, TR HJE, C AR, B AR; **88** T HJE, B HJE; **89** TR HJE, CR HJE, CL HJE, BL HJE; **90** TR HJE, CR HJE, CL HJE, BL HJE; **91** T HJE, B HJE; **92** T HJE, B HJE; **93** T HJE, C HJE, B HJE; **94** TR HJE, CR HJE, CL HJE, BL HJE; **95** TR HJE, CR HJE, CL AR, BL HJE; **96** T AR, B AR; **97** TR HJE, CR HJE, CL HJE, BL HJE; **98** T HJE, B HJE; **99** T HJE, C HJE, B HJE; **100** TR HJE, CR HJE, CL AR, BL HJE; **101** TR HJE, CR HJE, CL HJE, BL HJE; **102** TL HJE, TR AR, BL HJE, BR AR; **103** TL HJE, TR AR, B HJE; **104** TR HJE, CR HJE, BL AR, BR HJE; **105** TL HJE, TR HJE, B HJE; **106** T HJE, B HJE; **107** TR HJE, CR AR, BL HJE, BR AR; **108** TL HJE, TR HJE, B HJE; **109** TL HJE, TR HJE, B AR; **110** T HJE, B HJE; **111** T HJE, C AR, B HJE; **112** T HJE, BL HJE, BR HJE; **113** T HJE, B AR; **114** T HJE, B HJE; **115** T HJE, C HJE, B HJE; **116** T HJE, BL AR, BR AR; **117** T AGAMI Photo Agency/Alamy, BL HJE, BR AR; **118** TR HJE, CL HJE, BL HJE; **119** T HJE, C AR, B AR; **120** TR HJE, CR AR, CL AR, BL AR; **121** TR HJE, CR HJE, BL HJE, BR HJE; **122** T AR, B HJE; **123** T HJE, B HJE; **124** T HJE, B AR; **125** TL HJE, TR HJE, B AR; **126** TL HJE, TR AR, B HJE; **127** T HJE, B HJE; **128** T HJE, B AR; **129** T HJE, B HJE; **130** T AR, C HJE, B AR; **131** T HJE; **132** TL HJE, TR AR, B HJE; **133** T HJE, B AR; **134** T HJE, B AR; **135** T HJE, B HJE; **136** TR HJE, CR HJE, CL HJE, BL HJE; **137** TL HJE, TR HJE, B HJE; **138** TR HJE, CR AR, BL HJE, BR HJE; **139** T HJE, B AR; **140** T AR, B HJE; **141** T AR, B AR; **142** T AR, B AR; **143** T HJE, C HJE, B HJE; **144** T AR, B HJE; **145** T HJE, B HJE; **146** T HJE, C HJE, B HJE; **147** T HJE, B HJE; **148** TL AR, TR AR, B AR; **149** T HJE, B HJE; **150** T AR, B AR; **151** T HJE, B AR; **152** T HJE, BL AR, BR AR; **153** T HJE, C HJE, B HJE; **154** T HJE, B HJE; **155** AR; **156** T HJE, B HJE; **157** T HJE, B HJE; **158** T HJE, B HJE; **159** HJE; **160** TR AR, CR AR, BL AR, BR HJE; **161** TR HJE, CR HJE, BL HJE, BR HJE; **162** TR AR, CR HJE, B HJE; **163** T AR, B HJE; **164** T AR, B HJE; **165** T AR, B HJE; **166** T AR, B HJE; **167** T HJE, C HJE, B AR; **168** T HJE, B HJE; **169** T HJE, B HJE; **170** HJE; **171** T HJE, B AR; **172** T HJE, B HJE; **173** HJE; **174** T HJE, C HJE, B AR; **175** T HJE, C HJE, B AR; **176** HJE; **177** T HJE, C AR, B AR; **178** T HJE, B AR; **179** T HJE, B AR; **180** T HJE, B HJE; **181** HJE; **182** T AR, B HJE; **183** TR HJE, CR AR, CL AR, BL AR; **184** TL HJE, TR HJE, B HJE; **185** TR AR, CR HJE, CL AR, BL HJE; **186** TL AR, TR HJE, B HJE; **187** TR HJE, CR HJE, CL HJE, BL HJE; **188** T AR, C HJE, B AR; **189** T HJE, C HJE, B HJE; **190** T AR, B AR; **191** T HJE, B AR; **192** AR **193** AR; **194** T HJE, C HJE, B HJE; **195** T HJE, C HJE, B HJE; **196** T HJE, B HJE; **197** T HJE, B HJE; **198** T AR, B AR; **199** TR HJE, CR HJE, CL AR, BL HJE; **200** T AR, B HJE; **201** T HJE, C HJE, B HJE; **202** T HJE, B HJE; **203** T HJE; **204** T HJE, B HJE; **205** T HJE, B AR; **206** T HJE, B HJE; **207** T AR, B AR; **208** AR; **209** T HJE, C HJE, B HJE; **210** T HJE, B HJE; **211** TR HJE, CR AR, CL AR, BL AR; **212** T HJE, B HJE; **213** HJE; **214** T AR, C AR, B HJE; **215** T HJE, C HJE, B blickwinkel/Alamy Stock Photo; **216** T AR, B HJE; **217** T HJE, B HJE; **218** T AR, B HJE; **219** T HJE, C HJE, B HJE.

INDEX